PYTHEAS OF MASSALIA

Pytheas of Massalia (Marseille), mariner, explorer, geographer and astronomer, made a pioneering voyage into the then unknown Atlantic around 325 BC, reaching Britain and the Baltic; this book collects and translates the references to him and his book (which is lost), and discusses and explains them.

The Greeks of Pytheas' time knew virtually nothing of northern Europe beyond the often-fantastical stories of traders, and Pytheas was the first person to provide factual, first-hand information on this region. His journey covered Iberia, France, and Britain, the very existence of which he brought to the notice of the Mediterranean world; from where he travelled so far north that he encountered ice floes; he then reached the Baltic. It was he who recorded Thule, and his astronomy enabled him to locate it on the arctic circle. Two-thirds of our references to Pytheas come from Pliny and Strabo; their methods of work, as well as the perils of manuscript transmission, are explored in this volume. Scott also includes discussions and appendices on these areas to enable the scope of available references to be understood as a whole. There are some details of Pytheas' voyage that are lost, but the book offers balanced reasons for proposing how we may reasonably fill them in.

The breadth of Pytheas' achievements and the areas and topics his work covers mean that he has a wide range of appeal within classical studies and ancient history. This volume provides an invaluable resource to undergraduate and postgraduate students of early geography and astronomy, and Greece's knowledge of and relationship to the rest of Europe in this period.

Dr Lionel Scott is a classically educated retired barrister, whose publications include *Were There Polis Navies in Archaic Greece?*, BAR Int Series 899 (2000) 93–115, and *Historical Commentary on Herodotus Book VI* (2005).

ROUTLEDGE CLASSICAL TRANSLATIONS

Routledge Classical Translations provides scholars and students with accurate, modern translations of key texts that illuminate distinctive aspects of the classical world and come from a range of periods, from early Greece to the Byzantine empire. Volumes include thematic groupings of texts, texts from important authors as well as texts from the Byzantine period that are relevant for the study of the classical world but which remain inaccessible. Each volume has accompanying notes and commentary that provide a solid framework for deeper understanding of the material. As well as providing translations of significant texts, the series makes available material that is untranslated into English or difficult to access, and places these texts within new contexts to open-up areas of study and support research.

Titles include:

PLUTARCH'S THREE TREATISES ON ANIMALS
A Translation with Introductions and Commentary
Stephen T. Newmyer

MEGASTHENES' INDICA
A New Translation of the Fragments with Commentary
Richard Stoneman

THREE ANCIENT GEOGRAPHICAL TREATISES IN TRANSLATION
Hanno, the *King Nikomedes Periplous*, and Avienus
Duane W. Roller

PYTHEAS OF MASSALIA
Texts, Translation, and Commentary
Lionel Scott

For more information about this series, please visit: www.routledge.com/Routledge-Classical-Translations/book-series/CLTRA

PYTHEAS OF MASSALIA

Texts, Translation, and Commentary

Lionel Scott

Routledge
Taylor & Francis Group

LONDON AND NEW YORK

First published 2022
by Routledge
2 Park Square, Milton Park, Abingdon, Oxon OX14 4RN

and by Routledge
605 Third Avenue, New York, NY 10158

Routledge is an imprint of the Taylor & Francis Group, an informa business

© 2022 Lionel Scott

British Library Cataloguing-in-Publication Data
A catalogue record for this book is available from the British Library

Library of Congress Cataloging-in-Publication Data
Names: Scott, Lionel (Historian), author.
Title: Pytheas of Massalia : texts, translation, and commentary / Lionel Scott.
Description: Abingdon, Oxon ; New York, NY : Routledge, 2022. | Series:
 Routledge classical translations | Includes bibliographical references
 and index.
Identifiers: LCCN 2021023854 (print) | LCCN 2021023855 (ebook) |
 ISBN 9781032019987 (hardback) | ISBN 9781032020068 (paperback) |
 ISBN 9781003181392 (ebook)
Subjects: LCSH: Pytheas, of Massalia. | Geography, Ancient.
Classification: LCC G87.P92 S46 2022 (print) | LCC G87.P92 (ebook) |
 DDC 913/.36—dc23
LC record available at https://lccn.loc.gov/2021023854
LC ebook record available at https://lccn.loc.gov/2021023855

ISBN: 978-1-032-01998-7 (hbk)
ISBN: 978-1-032-02006-8 (pbk)
ISBN: 978-1-003-18139-2 (ebk)

DOI: 10.4324/9781003181392

To JFS, my wife

CONTENTS

MAPS

PREFACE

My interest in Pytheas goes back to a chance encounter with F30 some years ago. I have long been intrigued by the Greeks' attitude to science. They produced brilliant mathematicians and, to judge from Aristotle and Theophrastos, competent biologists; but where chemistry and physics were concerned, they chose dialectic and argument rather than experiment, though one must allow for the technical limitations of the time. Hero of Alexandria was a lone voice in the wilderness; there was no use for his steam engines, though we can recognise the social problems of replacing slave labour with machinery. I came across F30 at the same time as I was reading Galen (in translation), and was struck by the dichotomy: Pytheas spoke the truth about the frozen sea in the far north, and was disbelieved; Galen, a significant medical writer whose influence lasted some 1,500 years, quoted Homer and other poets as 'proof' of scientific facts. That sparked my interest in Pytheas, and this book is the result.

It is impossible to offer a commentary on the fragments of and references to Pytheas that will please everybody. A number of them do not name him, and one can argue whether they are close enough to his book to merit inclusion. In many cases, interpretation is controversial: where he went, what he saw, how well he understood what local non-Greek speaking informants told him, how he wrote it up, and how his text has been summarised, compressed, misunderstood or mis-recalled by later writers, often at second-hand. Any solution involves speculation and inference. Not everybody will agree with how I consider the balance of probability tips in a given case.

That having been said, and unless you dismiss his writings as fairy-tales, he was clearly a man of energy, intelligence, and knowledge; and he was the first Greek to visit what was, to the Greek world, the unknown north of Europe, including Britain, and then write about it. At first blush many of the references may seem unnecessarily long, with not much relating to Pytheas. I have so cited them (following previous commentators) to show the context of the Pytheas-related material, and to understand the working methods, and therefore reliability, of Pliny and Strabo. However much of Pytheas' work we can arguably recover, it must never be forgotten how much we have lost. What would we know of Dickens' *Bleak House*, for example, if all that survived was the statement that the first few pages described a cold damp fog penetrating every building in London? Not a complete parallel but a sobering thought.

Three final comments. Because Pytheas was the first to bring Britain to the notice of the Mediterranean world, he is of general interest to English-speaking readers; and while including references and citations for classical scholars, I have also explained some matters in more detail than might be necessary in a purely scholarly work. Secondly, his background must always be borne in mind. Whatever he did or failed to do, he must not be criticised for not having the knowledge or facilities of later generations. The third is that I have not attempted an exhaustive review of every previous commentator, or every paper that touches on some aspect of him. That is partly a time constraint, partly because a good deal of material is by continental writers, not readily accessible in Britain. Bianchetti's (1998) commentary may be consulted for its extensive bibliography.

This is not a critical edition as such, but textual problems that are relevant to understanding the substance of the Greek or Latin text are noted as they occur.

Lionel Scott
Bardsey, near Leeds
August 2021

ACKNOWLEDGEMENTS

I gratefully acknowledge the help I have received in putting this book together. The University of Leeds gave me honorary research status, and their librarians were always helpful when I needed assistance. Living close to the British Library at Boston Spa, I used their facilities, and I express particular gratitude to the staff for their help, and their patience and skill in always pointing me in the right direction. Professor Malcolm Heath provided several useful suggestions and pieces of information, as did Dr Roger Brock, who also read through parts of the book in draft and pointed me towards a number of references I might otherwise have missed. I owe particular gratitude to the late Professor Peter Rhodes, who very patiently and very acutely read through the whole work in an advanced stage of completion. My thoughts were sharpened and in some cases changed from the input when I gave a paper on Pytheas to the Leeds Branch of the Classical Association in 2017. I thank Dr Andrew Jones for pointing me to the scarf.scot website; and Lisa Benson and Professor Nils Anfinset of the Museum of Cultural History, Bergen, for providing copies of two Norwegian publications. I had particular help from two scientists, non-classical scholars but both interested in Pytheas: my wife, Dr June Scott, and my biologist friend, Dr Peter Hogarth (University of York). They gave invaluable help on details such as marine life, jellyfish and pearls from oysters, for instance, tides, the frozen sea, and amber and gems. Both also proved to be keen-eyed proof-readers. My wife also helped with the maps. My son Jonathan provided support on IT and computer issues, including help with the distances for the Coda. Routledge's reviewers made shrewd and helpful criticisms of various passages. Their editors Amy Davis-Poynter, Lizzie Risch and Marcia Adams made useful comments and were always patient with me; I also thank my copy-editors, Denise File for her patient help and Brian Frascati for useful suggestions and correcting misspellings. Finally, I thank De Gruyters for permission to use the texts of their Teubner editions as follows: the Greek text of F2 from Cleomedis Caelestia, ed R Todd (Leipzig 1990), and the Latin texts of F10 and F11 from Martianus Capella, ed A Dick, J Préaux (Stuttgart 1969, 1978).

It goes without saying that the errors which have escaped others' scrutiny are my own, including what I might term the quasi-errors, where gaps in our knowledge of Pytheas and his book can only be filled in, or partly filled in, by inference and guesswork. I cannot hope that all my solutions will appeal to all my readers. Also, I cannot guard against H G Wells' don who 'lies in wait to wreak his vengeance on the footnotes of the book he had hoped to write' (quoted by R E Megarry, *A Second Miscellany-at-Law* (London 1973) 244). In strict logic, it is impossible to try to see the world through Pytheas' eyes as he went from place to place and then wrote it up, but the effort of trying to do so by going behind the mere words of our texts is rewarding and well worthwhile.

ABBREVIATIONS AND CITATIONS, ETC

Unless otherwise stated, all dates herein are BC.

For the spelling of Greek masculine names, I retain their -os terminations, unless I am citing them from a Latin text. In some cases, I retain the original -ai- diphthong, but use the Latin -ae- where that form is more familiar; I retain the more familiar 'c' as opposed to 'k': so Phocaea, Dicaiarchos (not Phokaia, Dikaiarkhos); I also retain the more familiar Byzantium (strictly 'Byzantion'). Authors are cited in line with the conventions of Liddell and Scott for Greek texts and the Oxford Latin Dictionary for Latin.

An important aspect of the Greek and Roman view of the world was that most believed that the Europe-Asia-Africa landmass was surrounded by sea, the 'ocean' in both languages. I have used 'Ocean', with capital 'O', in this sense, using 'Atlantic', 'North Sea', etc for clarity where a particular part of the Ocean is involved. In general, modern place names are used where no confusion can arise, but their ancient equivalents where it makes for better clarity or we need to see the place through Greek or Roman eyes: e.g. France, or Keltikē, the land of the Celts (which was basically France and Belgium); later Galatia (Greek) or Gaul (Latin); Marseille, or Massalia (Greek) or Massilia (Latin); Spain, or Iberia.

Abbreviations: except as noted in the Bibliography, reference works and journals are cited as in *L'Année Philologique* and other standard bibliographical works.

Unless otherwise stated, all translations are my own; I acknowledge my debt to existing ones which I have consulted, in particular where I have copied a particularly felicitous phrase; many sentences can only be sensibly translated one way.

Fragments are cited in line with standard editions; in particular Eratosthenes from Roller (2010), Hipparchos from Dicks (1960), and Posidonios from Edelstein and Kidd (1988–1999).

Modern latitudes are from the Wikipedia entry for the place; summer solstice daylight figures and summer and winter solstice sun heights are extracted from timeanddate.com.

DISTANCES AND TIMINGS

This Note deals with the various questions of distances and timings that arise.

(1) Terrestrial distances in stades: the basic Greek unit of length was the stade, 600 feet; but the foot varied from place to place, between 11.64″ and 13.1″.[1] Thus the stadium at Olympia, 600 feet, is 210 yards (192.28 m) long, corresponding to a foot of about 12.6″. But all our texts refer to long distances. None were surveyed, none reflect the difference between one city's foot and another, and all are expressed in round figures. A report that polis A was 1,000 stades from polis B would typically be a traveller multiplying a journey of x days by his estimate of covering y stades per day, rounded up or down: generally see Roller (2010) 271–3. Pothecary (1995) 49–50 correctly proposed that Greek geographers thought in terms of a standard stade, but this would apply equally to their oral sources. Nonetheless, there are several discussions as to which foot we should use in turning stades into miles, e.g. Thomson (1948) 161–2, Dicks (1960) 42–6. The reality is otherwise: I have treated the foot as 12″ and rounded the result up or down to the nearest 5 miles. The kilometre equivalents are similarly rounded up or down. The result gives the reader a perfectly good picture of what our text means. In this sense, the stade was a concept, not a measured length.[2]

(2) Roman miles: the Roman mile was equivalent to 1,481.5 m (1,620 yards).[3] The usual Roman translation of stades was at 8 to the (Roman) mile (Strabo 7.7.4).[4]

---

1 See, e.g., *OCD* sv Measures.
2 So F29, Hipparchos' distances were based on one degree of latitude being 700 stades; as there noted, that depended on Eratosthenes' 252,000 stades for the size of the earth. Both men were thinking of a 'standard' stade.
3 The Roman mile was 1,000 paces, and the *passus*, pace, meant two steps totalling five feet. These cannot be translated precisely, their foot being 29.5 or 29.6 cm, 11.6″ or 11.7″: see *BNP* sv Measures; *OCD* sv Weights and Measures.
4 Strabo adds that Polybios used a multiplier of 8⅓; but by Strabo's time it is probable that a multiplier of 8 was generally used.

(3) Actual land distances: distances by road between two places have been obtained using Google and Bing maps; distances expressed as 'as the crow flies' have been measured either from them or with a ruler on paper maps; both should be taken as subject to a small margin of error.

(4) Ancient maritime distances have similar problems to land distances: they are mariners' estimates, turning the length of the voyage into stades. Maritime communities no doubt had their own rules of thumb for this, though the time of a particular voyage would depend on winds and weather and whether one could use the open sea or kept close to the coast. At some point, commonly accepted distances between ports were recorded in literature; they must be read accordingly. The limits of accuracy are illustrated at Strabo 2.5.24: from Rhodes to Alexandria 'with a north wind' the distance is 4,000 stades (455 miles, 730 km); others report 5,000 stades (570 miles, 910 km); Eratosthenes using *gnomon* readings (Appendix 2 §3) calculated it at 3,750 stades (425 miles, 680 km). The modern distance is 330 nautical miles (nm) (380 miles, 610 km).

(5) Marcianus, a page further on from F9 and still within his introduction, says that 700 stades per day[time] is agreed with a following wind; he associates better or worse times with how well the ship is constructed. An overview of the literature suggests that a typical rule of thumb was in the region of 700 stades (80 miles, 130 km) in a day, and 1,000 stades (115 miles, 185 km) for a day and night: see Arnaud (2005) 70–81.[5] Lists of the speeds of known voyages are a useful guide, though they cover ships of different types, and some of the Roman ships had banks of rowers to supplement wind power. Casson (1996) 283–8 lists those with favourable winds, and at 289 with unfavourable. The range of the first is 2.8 to 6.2 knots (*op cit* 283–8), equivalent to 3.22 to 7.13 mph (5.19 to 11.48 km/h); the second, 1.5 to 3.3 knots, 1.73 to 3.8 mph (2.78 to 6.11 km/h).[6] The voyages were of continuous days and nights, and presumably under men who knew the route; but they are a useful guide to what in practice

5 Another example is Herodotos' statement at 4.86.1, that a day's sailing is usually stated to be approximately 700 stades (80 miles, 130 km), and at night 600 stades (70 miles, 110 km). He actually expresses this in fathoms (100 fathoms = 1 stade). It is problematic on two counts: (1) he uses the figures to offer a number of distances between places which are not accurate: see Asheri *et al* (2004) 642-3. (2) He says that this is the case in high summer, using the *hapax* μακρημερία, *makrēmeria*, 'long day'. Arnaud (2005) 78–9, treats the passage as referring to the actual solstice, with a day to night ratio of 17:7. However, in the Mediterranean, the ratio of day to night is some 15:9; nights are not much shorter than days in the earlier and later months of the sailing season. On the other hand, as noted for Pytheas below, neither Herodotos nor his sailor informants had equipment for measuring these periods of time.

6 He also lists speeds for fleets; this is not relevant here, as a fleet would usually take longer than an individual ship.

was achievable. Arnaud's tables are at *op cit* 102–4, with a range of 1.6 to 7 nm, 1.84 to 8.05 mph (2.96 to 12.96 km/h).[7]

(6) Our sources behind §§4 and 5 reflect sailings in the Mediterranean; it is here assumed that in the sort of boat Pytheas probably had (pp 165–6) his overall rate of progress in the Atlantic would not be significantly different. We are interested both in the sort of speeds and progress we may attribute to him, and how he came to state the dimensions of the coasts of Britain, as recorded in F4. We may assume that he had good previous maritime experience; he would inherit whatever rule of thumb prevailed in Marseille for turning the time of a voyage into stades. Once past Gibraltar, the coasts would be unfamiliar to him and his crew. He would have the skill to read an unfamiliar coast; even so, he would want to keep closer to the coast, and so make slower progress, than someone familiar with the route and able to keep more to the open sea. But 'open sea' is relative; the typical location of ancient wrecks suggests mariners typically kept headlands in sight and did not venture very far into open waters on a coastal route; cf F9 n 2.[8] Thus on a good day with favourable currents and a fair wind, Pytheas could achieve 3 to 4 knots, say 4 to 5 miles per hour, so at least 70 miles, 110 km, and close to 700 stades. In the estimates of his progress in the Coda, pp 170–7, much gentler progress is assumed. His dimensions for the sides of Britain must be based on his rate of progress; as noted in the Coda p 174, he had no means of measuring his progress mechanically, whether hours, speeds, or distances; see p 32 and pp 174–6 for how accurate his dimensions are.

(7) Actual sea distances have been obtained in two ways. Those for the Coda have been obtained using the 'Distance Calculator' tool at www.sea-seek.com, on the basis of a ship keeping within sight of the shore and using headlands where possible, but also taking into account the need to negotiate difficult stretches such as the bays up the west coast of Britain, the Scottish islands, and around the Danish islands. Others, such as that from Boulogne in F18, are a direct measurement on a map. All the distances should be regarded as realistic but ballpark figures and treated accordingly; the reader who makes his or her own estimates of the distances may well come up with other figures, but they are not likely to be significantly different. The number of days allowed reflect the need to put in to land periodically, e.g. for water and supplies.

---

7 The distance and time in Strabo 4.3.4 is problematic: 'from the rivers of Keltikē to Britain is 320 stades; if you leave on the evening ebb-tide you arrive at the eighth hour next day'. That would be 36 miles, 58 km, in about 16 to 18 hours, a little over 2 mph or 3.6 km/h. Which rivers? None of his usual passages to Britain, F38 n 1, are anywhere near 36 miles away. It would suit Sangatte or Wissant to Richborough, recently used by Caesar (cf F18 n 1), though there is no river at either. Boulogne is on a river, the Liane, and is about 46 miles, 74 km from Richborough. Perhaps Strabo reflects one of Caesar's crossings; as in n 6, a fleet moves more slowly than one vessel.

8 Overnight voyages between, say, Keltikē and Libya, or (later) from Rome to Alexandria, required different experience and skills.

MINI-BIOGRAPHIES

Ancient authors who are named just once or twice are noted in the commentaries as their names occur, but others appear regularly, and it is convenient to offer brief biographies here.[1]

Artemidoros of Ephesos was a first century geographer about a generation older than Strabo, who often used him. It would appear from F34 that he, like Strabo, did not approve of Pytheas.

Dicaiarchos, born c 375, was a pupil of Aristotle who wrote on a wide range of subjects, including geography. One of his ideas is of particular interest to us: he defined a line of latitude that ran from Gibraltar through Rhodes and on to the Himalayas, which was very influential for subsequent geographical writers.

Diodoros, commonly referred to by his Latin name of Diodorus Siculus, was born in Agyrium (now Agira) in Sicily in the earlier part of the first century; he lived largely in Rome and wrote a 'universal history' in 40 books. It was essentially an epitome of various earlier writers. Only some survives in full, including book 5. Our F4-F6 comes from a section in book 5 dealing with Britain and Baltic amber; here he epitomised Timaios (below), who in turn was using Pytheas.

Eratosthenes of Cyrene, third century, was a scientist of many abilities: mathematician, astronomer, philologist, chronologist, to whom others attached the label 'beta', as if to suggest that he tried to excel in too many disciplines. He was also librarian at Alexandria. He is perhaps best known for the major achievement of measuring the circumference of the earth, by noting that at the summer solstice, the sun at noon was directly overhead at Syene (Aswan), shining directly into a well, whereas at Alexandria it cast a shadow. Treating the one as due north of the other and using the accepted distance between

---

1 Fuller biographies are readily available in standard reference books such as *OCD* and *BNP*, and also in Keyser and Irby-Massie (2008). For those whose texts survive more or less complete, acceptable English translations are generally available online, though the notes to some are often now out of date.

them, 5,000 stades (570 miles, 910 km), and assuming that the earth was a perfect sphere, he calculated the circumference of the earth as 252,000 stades (28,635 miles, 45,815 km).[2] One of his works was a 3-volume *Geographika*, geography, which either included a map of the world, or at least explained how to draw one. He is credited with inventing the word 'geography'; certainly he is the first author known to have used the word.[3] For northern Europe, he drew on Pytheas, still the only source for that region. For some parts of the world, especially the east, he had to use distances which were not always very accurate, creating problems for later geographers.

Herodotos of Halicarnassos (now Bodrum, Turkey) was the first to write history in the sense of real events, as opposed to the mythical past; he was born c 485–480 and died c 425 or perhaps a little later. His world was essentially without archives, but he travelled widely and tried to find out what people had to say about past events and their background, and his work is a treasure-house of the oral traditions of the time.

Hipparchos of Nicaea (now Iznik, north-west Turkey) was a very important astronomer and geographer of the second half of the second century. He had the advantage over Eratosthenes in two respects: for some areas of the world more accurate information, though for northern Europe probably only Pytheas; for astronomy he had knowledge of trigonometry. He produced a table or tables of latitudes with the various lengths of daylight at the summer solstice; and his latitudes influenced his geography. Only the astronomical work from which F8 comes survives in full.

Philemon was a Greek geographer of the early first century AD. Only four fragments survive. One is in Ptolemy's *Geographia* at 1.11.7, the dimensions of Ireland he got from merchants.[4] The others are in Pliny: one in F17, about the Cimbri on Jutland, and two at *NH* 37.33 and 36 about amber in Scythia (just before and just after F21: see on F21 nn 2, 8). It is tempting to suggest that he had all these details about the north because he had been there; Tierney (1976) 260–2 explores the possibility that Philemon did so, at least as far as Ireland, which was Tierney's interest. But he cannot have gone into the Baltic: it is difficult to think that in the first century AD he could speak of amber coming from Scythia rather than where German tribes live.

---

2 His basic calculation was that at Alexandria, the angle between the *gnomon* (vertical stick or rod) and the path of the sun's rays as ascertained from the shadow they cast was one-fiftieth of four right angles: Heath (1921) I 106-7; Roller (2010) 23, 263-7. Alexandria is slightly north-west of Aswan, but for its time the accuracy is remarkable. Modern measurements are: equatorial circumference 40,076 km, 24,902 miles; polar circumference 40,007 km, 24,859 miles.

3 From γῆ, *gē*, 'earth' or 'world', and γράφω, *graphō*, 'I write'.

4 Strictly, it was Ptolemy's predecessor Marinus of Tyre (F18 n 1) who recorded Philemon, and Ptolemy copied Marinus.

Pliny the elder, 23–79 AD, is well known for taking a squadron of ships when Vesuvius erupted to try to rescue those fleeing Pompeii from the beach at nearby Stabiae, and dying in the attempt. He combined a military and legal career with extensive reading, leading to a considerable literary output, including a 37 volume *Natural History*, a compendium of knowledge based on this reading. His method of working and the value of his work is discussed on F14.

Posidonios of Apamea in Syria (now Qalaat Al-Madiq), c 135–c 51, was a Stoic and distinguished scientist who wrote on history, geography, and astronomy. He travelled in the Mediterranean and past Gilbraltar as far as Cadiz, but it is very doubtful that he went further west.[5] His work is lost, and we know of him through later texts, including Strabo. His writings included an *On the Ocean*, which is also the title of Pytheas' work; it is unclear if he used him: see p 2.

Strabo of Amaseia (now Amasya, Turkey), c 69 BC–21 AD, lived for a time in Egypt, but mostly in Rome. He wrote a *History*, mostly lost, and a *Geography* in 17 books. We owe him about half our references to Pytheas, nearly always with a pejorative comment. As noted in the introduction to F24, he had a rigid Stoic view of the world, leading him to assert that anywhere north of Ireland was uninhabitable because of the cold; what Pytheas said about the far north was therefore lies, meaning in turn that everything Pytheas wrote about the north was lies. It is unclear if he had read Pytheas' actual book; it is usually obvious that he refers to Pytheas when discussing an intermediate author, Eratosthenes and Hipparchos in particular, who had used Pytheas for some detail. Dion (1965) 450–2 argues an alternative reason for Strabo's hostility to Pytheas: Pytheas offended the amour-propre of the Romans, whose cause Strabo's family supported. He wanted to downgrade Pytheas, who had visited 'all Britain' (F30), when the distinguished Roman Caesar had failed to conquer it. His general approach is reflected in his stated wish to make his work suitable for statesmen and practical men and not trouble them with scientific detail: 2.5.34, on which see Dicks (1960) 171, 187. This perhaps hides his own difficulty with mathematics, noted in some of his references.

Timaios of Tauromenium (now Taormina, Sicily), c 350–c 260, and so perhaps a little younger than Pytheas, lived most of his life in Athens. He wrote a comprehensive work entitled 'Sicilian Histories' but in fact ranging much wider and including topics such as geography and ethnology. His work survives only in fragments; these include F4–6, Diodoros' epitome noted above, dealing with Britain and Baltic amber. Because he drew on Pytheas, he is important for confirming the latter's dates (Coda p 167).

---

5 See Kidd (1988) 16-21; Posidonios' observations of the apparent increased size of the setting sun appear to have been made at Cadiz and not at the Sacred Cape (*ibid* 463–4, on F119 E–K, Strabo 3.1.5).

INTRODUCTION

1 Pytheas and his book

Pytheas was a native of Μασσαλία, *Massalia* (Massilia in Latin texts), Marseille, and was the first Greek to sail into the north Atlantic and visit Britain. He then wrote a book about his voyage, which unfortunately is lost; but there are some 30 references to him in surviving classical literature, and several other passages which reflect what he wrote.

There are questions as to his dates, but for the reasons canvassed in the Coda, pp 166–8, his voyage almost certainly took place in the third quarter of the fourth century, most likely between c 330 and c 325, with his book being published a year or two later.

Each citation is given a conventional F for fragment number, but they are called 'references' in this book. Strictly speaking, a fragment is a quotation of another author (e.g. the lines of poetry in F34), or at least a fair summary of the other writer's words. But how far surviving texts can be said to contain something of an earlier writer's work is never easy to define; much depends on what the fragment collector is aiming for.[1] My aim has been to try and recover as much as possible of Pytheas' book, and behind that, to identify the route of his voyage. The decision what to include and exclude is not always easy. Except for Pliny, it is doubtful whether any of our texts were written directly from Pytheas' book, so his text mostly survives at second- or third-hand. Eratosthenes and probably Hipparchos did read him; but they extracted the details they wanted, not necessarily quoting him verbatim. But they in turn survive only as fragments in Strabo, who criticises them for relying on Pytheas; we cannot assume that he in turn cited them verbatim. As noted in the Preface, it is difficult to decide what texts to include where Pytheas is not mentioned by name; in theory, Virgil's 'ultima Thule', noted below, is a fragment. It was tempting to preface each citation with R, for 'reference', because even when he is named, our references give the substance of what Pytheas wrote, not his actual words.

---

1 See the papers in Most (1997).

DOI: 10.4324/9781003181392-1

It is clear that his book made an immediate impact: it brought Britain and northern Europe to the notice of the Mediterranean world. His younger contemporaries Timaios used him (F4 to F6), and Dicaiarchos read him (F30). Apart from Eratosthenes and Hipparchos, a number of other authors may have used him. There is a very good argument that the author of the *Prometheus Vinctus* attributed to Aeschylos used Pytheas for his European geography: Finkelberg (1998) at 131.[2] Crates of Mallos, a grammarian of the first half of the second century, may have used his mention of very short nights in the far north, F7, to prove that Homer knew of this when speaking of the Laestrygonians.[3] It is likely that Posidonios (see p xxi) read him, but it is hard to find clear evidence that he used him. His principal work of geography was entitled *Peri Oceanou* (About (the) Ocean), like Pytheas' (F3, F7),[4] and he may have thought that Pytheas' title suited his own work ('can be traced back to Pytheas', Kidd (1988) 219, following Mette (1952) iii, 13–14, 36–8). Mette also suggested that Posidonios was the source for Strabo's references to Pytheas' abilities in astronomy and mathematics, F38 and F39. Like Pytheas, Posidonios argued that the moon caused tides (see on F1), and he may have noted what Pytheas wrote to confirm his own deductions; but his thinking on the Ocean and tides covered wider and different ground to Pytheas (F214 to F220 E–K).[5] Tierney (1959–60), (1976), arguing that Pytheas knew Ireland, sought to attribute much of Posidonios' Celtic ethnology to him, e.g. (1976) 260, 'We know that Posidonius read and appreciated Pytheas'. But all that Pytheas may have known about Ireland was from seeing the Antrim mountains and thinking that they were on a small island on his port side. If he did record it, it is lost: pp 181–2. As to Posidonios, there are serious questions as to how much he actually wrote about the Celts: Nash (1976), Kidd (1988) 308. Posidonios refers to tin from Britain going to Marseille, F239 E–K = Strabo 3.2.9; Roller (2018) 139, noting that Strabo's best MSS spell the island *Prettania*, suggests that this might go back to Pytheas (cf on F5). Strabo, who calls Posidonios 'the most learned (πολυμαθέστατος, *polymathestatos*) scientist of my time', never criticises him for using Pytheas.[6] In late antiquity, there are texts with some dependence on

2 Her closely reasoned argument carefully analyses the text of the play and other sources to reach this conclusion. She proposes a fourth, not fifth, century date for the play, which is not universally accepted.
3 The Laestrygonians, who lived where there are long days and short nights, are at *Od* 10.79–85. The main passage is Gem 6.10–11, which immediately follows the mention of Pytheas F7; and see Evans and Berggren (2006) 163; Broggiato (c 2001) 217–18; cf p 2.
4 F49 E–K = Strabo 2.3.1.
5 Kidd (1988) 463, on a passage of Posidonios dealing with reports that in the Ocean the sun became larger when setting and made a sizzling noise, says that 'they probably go back to Pytheas'. This is puzzling, because he cites F7, which relates to Pytheas and the Northern Isles, and Tac *Germ* 45.1, about the Baltic, and nothing to do with Pytheas. The sizzling story sounds like folklore; but the setting sun can sometimes seem larger than usual, especially on coasts, and Pytheas might have noted it. Posidonios had only been as far west as Cadiz. Posidonios' own thinking about the arctic circle might go back to Pytheas: see on F2 n 6.
6 Roseman (1994) 18–19 has a useful note on Crates' and Posidonios' possible knowledge of Pytheas.

Pytheas, but they are so general, copied from earlier writers, and guesswork, that they add nothing useful; see, for instance, the references to Dicuil and Procopios at Appendix 6 §13. As a footnote, it is not clear for how long copies of his work survived; see on F9.

Two of Pytheas' details, Thule and the frozen sea, entered popular lore, though divorced from both his name and their contexts, that Thule was on the arctic circle and six days' sail from Britain, and the frozen sea even further north. Virgil could only speak of 'ultima Thule' in his *Georgics* (1.30), published c 29, if people already knew of it as a distant and remote island; so Seneca in the next century (*Medea* 379).[7] Around the same time as Virgil, Varro, noting different climates in his book on agriculture, wrote that in the far north they say that it is not possible to sail in the ocean because of the frozen sea (*RR* 1.2.4). In the early first century AD, Juvenal, *Sat* 2.1–2, spoke of the 'glacialem Oceanum', 'frozen sea', in the distant north.[8]

A cursory glance at our references shows that Pytheas was an intelligent man of many parts: navigator, explorer, enquirer, geographer and astronomer. Each of the references will contribute something towards building up as complete a picture of the man and his work as is possible.

2 Problems of presentation

There is no ideal way to list the references. A chronological list is impracticable, since several authors themselves only survive as fragments: Timaios, Eratosthenes, and much Hipparchos. Roseman lists surviving authors chronologically, but divides up many texts in an attempt to distinguish mentions of Pytheas, *testimonia*, and actual citations.[9] Bianchetti (1998) lists by topic, but many references touch on more than one topic; in many cases, one reference has to be considered in conjunction with another. Here, the references are listed alphabetically by author, with cross-references and Appendices for common matters. The whole contains several details of general interest; to keep the book within reasonable limits, some discussions have been compressed; one can only apologise if the reader would have liked to know more.

There are several texts where the spelling of foreign names for places and tribes is problematic. It should be self-evident, though not often clearly stated, that a foreign name, often from an illiterate place and with no local spelling to help, could only be written (and pronounced) in Greek or Latin within the limitations of that alphabet, and the consonants and vowels available; Greek and Latin pronunciation would typically differ to that where the name originated. Some of Pytheas' names were heard by him as

---

7 Mela, F13, reports Thule in Greek poetry, but nothing survives. For Thule in later lore, see Appendix 6 §13.
8 Although Varro had referred to Eratosthenes' *Geography* a few lines earlier, the context of his discussion of different climates is on the basis of popular knowledge, not a written text.
9 How closely the texts in her citations reflect Pytheas' actual words is a question.

pronounced by the locals when he was there. Usually, foreign names circulated orally in the Mediterranean, typically from merchants and mariners speaking of the origin of their goods, and also, for Germany, by army personnel who had served there; we know of them only because they eventually got recorded. Thus the same name might be pronounced differently by different people, and the eventual spelling reflect that. We should not assume that there was always a standard spelling of a foreign name.[10]

There are also the perils of transmission. In Roman times, one method of publication was by dictating the text to a roomful of scribes; a man under the pressure of keeping up could spell an unusual name differently to that in the master copy. Making a one to one copy, as became the norm later, was also liable to create a misspelling, for instance from confusion between different letters (I and T, for example), and more likely with an unfamiliar name.[11] In the case of Pliny, who used dictated notes, including shorthand notes (see on F14), both their making and transcription could produce mistakes. It is not always possible to be certain how an author originally actually spelled a name, and behind him, how closely that corresponded to how the locals themselves pronounced it. Some of these problems are noted as they occur; the particular problems of German tribes are discussed in the Note to Appendix 4, and those of a tribe in Brittany in Appendix 5.[12]

The Greek for the inhabitants of any non-Greek place was βάρβαροι, *barbaroi*, which in Greek meant only those who did not speak Greek, with the underlying implication that they also lacked the other advantages of being Greek. I have translated this as 'locals', and use the term generally for the inhabitants of a non-Greek place. 'Barbarians' now has a different connotation, and 'natives' can imply a more primitive way of life than would be appropriate. Greek used πόλις, *polis*, for almost any town or settlement, Greek or non-Greek; I use the conventional translation 'city', but it could refer (for instance) to the handful of buildings around a harbour or trading post.

---

10 Note that (a) Pytheas' original spelling would also depend on how he recalled the name when writing his book, unless we assume that he made notes, say on wax tablets, as he went; and (b) where a Latin text depended on Greek originals, it was usual to spell it with its Greek declension, particularly feminine nouns.

11 A decision on the 'correct' spelling may be impossible. Consider, for example, the gentleman mentioned in Caes *BG* 1.19.3: did Caesar spell his name Procillus or Troucillus (or something else?). Another example, closer to home, is that the Grampian mountains are so called because of a misreading of 'Graupium' in a manuscript of Tac *Ag* 29.2; it duly appeared as 'Grampium' in a printed edition of c 1475 AD.

12 So far as we know, the Bretons, certainly in Pytheas' time, were illiterate, so there was no local spelling to assist; if they already had coinage, it was just copies of Greek coins. Germanic tribes were always illiterate. The problem of 'correct' spelling can be assessed where the original is from a literate society, the Greek spelling of Persian names in Herodotos: so Agbatana (later Ecbatana)/Hagmatāna; Bactra/Bāxtriš; Armenia/Arminiya; Smerdis/Bṛdiya (Schmitt (1967) 123–4, 121). Different men called Megabazos/Megabyzos and Megabyzos/Megabyxos appear 43 times in Herodotos, with both MS variants and also whether it is the same name (Bagabuxša) or two different ones (*ib* 121 with n 17). A modern comparable is the different spellings and pronunciations of Bombay/Mumbai and Pekin/Beijing.

THE BACKGROUND TO
PYTHEAS' VOYAGE

Any assessment as to why Pytheas made his expedition, where he went, what he learnt, and what he wrote, has to be made against his background, education, and perception of the world; and also the political circumstances of the time. We may start with Marseille, which is sufficiently proud of him (and Euthymenes, referred to below) to have statues of both outside their Palais de la Bourse.

1 Massalia, trading city

Any Massiliot was heir to a long maritime and trading tradition. She had been founded c 600 by Phocaea, a city on the west coast of Ionia (Asia Minor), now Foça.[1] Phocaea was a strong maritime state, noted for its extensive trade links, and unusual in using penteconters rather than merchant galleys.[2] She located an ideal spot for what would initially be a trading post:[3] a bay where an excellent harbour could be made, with good access to and from the interior, and half a day's sail (some 35 miles, 56 km) from the mouth of the river Rhone. She soon flourished, and archaeology confirms that she became an important trading hub for a variety of commodities. She also founded a number of daughter settlements both

1 It is possible that there was some resistance from Carthage, but the reference is ambiguous: see p 17 n 16.
2 Hdt 1.163.1 noted both her extensive trading partners: the Adriatic, Etruria, and different parts of Spain, and her use of penteconters. The penteconter ('50 oarer') was commonly a troop carrier or warship. It was thus a long ship; powered with both sails and oars, it would always be better at not being at the mercy of adverse winds and avoiding a pirate ship than would the typical merchant galley, which relied mainly on sails. The typical merchant galley was shorter and broader ('round ship'), and had the advantage of carrying larger loads. We should not assume that a penteconter had to have 25 oars per side; it was probably a generic word and could apply to a ship with (say) 42 or 44 rowers. In trading, that number would make little difference to speed or the size of its load, and entail fewer men to feed and pay.
3 Hansen and Nielsen (2004) 165–7. Boardman (1999) 162 rightly stresses that many foundations were initially as much for trade as to reduce population pressures. For the foundation, see also *ibid* 214–15; *CAH*[2] III.3 140–1 (Graham); Malkin (1987) 69–72, with stress on its religious aspects.

DOI: 10.4324/9781003181392-2   5

east and west. Those to the east do not touch on Pytheas;[4] to the west, there were Emporion (Empúries, Spain), founded c 600;[5] and a generation or so later Agathē (Agde, France). In addition, she established a number of other trading posts or settlements down the coast of Iberia, the most important being Hemeroscopeion and Rhode, the furthest (probably) being Mainakē.[6] There was some diminution of trade c 500. In so far as this was connected to disturbances in mainland Europe associated with the transition from Hallstatt to La Tène cultures some three or four generations before Pytheas' time, it seems to have soon recovered.[7] Cunliffe (2001) 12–18 and Roseman (1994) 152–5 have helpful summaries of the sorts of goods involved, exports in local wines and fruits, for instance.

Two commodities in particular, both valuable in the ancient world, passed through Marseille: tin and amber. We must distinguish their actual origins and what Pytheas and his contemporaries believed. Amber came from the west coast of Jutland, and Samland, the Gdansk area. It will have been the former that reached Marseille; the latter went to the Adriatic: see further on F6. Pytheas will have known only that it reached Marseille from somewhere in the north, probably with some of the stories of its origins that Pliny would record.[8] Tin reached Marseille from Cornwall, Brittany, and Spain. By Pytheas' time, the latter possibly came in Carthaginian ships, but the rest came overland, as noted on F5 §4. He would believe that it all came from the tin islands, an archipelago said to exist somewhere in the Ocean in the distant north. As noted at pp 8–9, they were a traders' creation to disguise their true origin. There is an important detail to all this trade: it is

4 The most important are Tauroention and Olbia (Le Brusc and Hyères, west and east of Toulon), and Nicaea and Antipolis (Nice and Antibes) (Strabo 4.1.5, 9).
5 Possibly in conjunction with Phocaea itself.
6 For these places see Hansen and Nielsen (2004) 161–8. Strabo 3.4.6 speaks of three small cities including Hemeroscopeion. Except for Emporion and Rhode, Ciutadella des Roses, there is uncertainty about locations. Suggestions are Hemeroscopeion, Denia; one of the two others could be the island Alonis of Herod *Pros Cath* 3.1.96; it is variously located at Villejoyosa, north-east of Alicante, or S. Pola or Guardamar del Segura, to the south-west. The other could be the place later settled by Hamilcar as Acra Leuce ('White Promontory', Diod 25.3.10), at or near Alicante. Mainakē was perhaps near Torre del Mar, some 12 miles east of Malaga; but it is also suggested that the texts which speak of Mainakē as a Massiliot colony were wrong, and that it was not a Greek foundation: Aubet (2005), esp 191–3. For Pyrene and Cypsela, see Hansen and Nielsen 162.
7 Boardman (1999) 217–19, 224; Roseman (1994) 152–5; Cunliffe (2001) 14–18. The extent of the disturbance to Massiliot trade by aggressive Celtic migrations or invasions is contentious, as is the history of how Celts came to occupy much of western Europe. They were certainly established in some areas before 500, as both Hecataios *FGrH* 1 F18a and Hdt 2.33.3 and 4.49.3 show. Discussions such as *BNP* sv Celts and the Wikipedia article en.wikipedia.org/wiki/Celts show that there is broad agreement as to evaluating the evidence but differences as to the detail. There is general agreement that by Pytheas' time Massalia was heavily involved as an exchange point for goods coming through France from the north for onward shipment to the Mediterranean, and vice versa.
8 The trade was considerable, though not all reached Marseille: it is commonly found in graves in parts of western Europe from around the sixth century: see, for instance, *BNP* sv amber ('Celtic inhumations of the late Haltstatt and early La Tène period'); cf Cunliffe (2001) 17. See on F21 for the stories about amber.

likely that Pytheas, in common with many Massiliots, had some knowledge of the local Celtic dialect, both from those carrying goods into and out of Marseille, and (probably) the Celtic population in the city itself. This would help him communicate with the locals in France and Britain.

We know nothing about Pytheas' background and family, except that they were able to give him a good education (see section 5). But he could not have undertaken his voyage unless he also had fair experience as a mariner. If his family were not part of the local trading community, they must have known many who were. It is generally assumed that one reason for his voyage was to investigate the origins of tin and amber, and see if he could source them more cheaply. It is easy to see that they would offer Pytheas help in being provided with a suitable ship: well built, and oared by a small crew, enough to give power if there was no wind, and not too many mouths to feed.[9] They could further help with supplies of wine and other commodities for Pytheas to use to trade *en route* for supplies and water, and generally to help pay his way.[10] Perhaps he was also able to take some silver bullion for such payments.

2 Knowledge of the north

Few solid facts about the north, as opposed to stories, were known in Pytheas' time. In c 500 Hecataios of Miletos published his 'world tour'.[1] Book 1 covered Europe, book 2 Asia and Africa.[2] Some 150 names survive from book 1, but they are all on or near the Mediterranean and Black Sea coasts. He had almost no information for central or northern Europe; Britain was unknown; at most, he knew that Keltikē, the land of the Celts, lay north of Marseille.[3] From c 450, Herodotos wrote his *History*, a more wide ranging work than its modern title suggests. He travelled the Mediterranean and Black Sea, and perhaps further east, though probably no further west than southern Italy, enquiring as he went. His book contains

---

9 Hawkes (1977) 44 suggests that his crew might include Celtic speakers, which is quite feasible; more imaginative is his addition of a Scythic speaker in the crew. See on F7 for a probable reference to his crew, and pp 166–7 for his boat.

10 Dr Roger Brock suggested this as a realistic possibility.

1 Our sources offer two titles with a similar connotation: Περιήγησις Γῆς (*Periēgesis Gēs*, 'Guide round the World') and Περίοδος Γῆς (*Periodos Gēs*, 'Journey round the World'; similar works were also called Περίπλους, *Periplus*, 'Sailing round' or 'Circumnavigation'. In Hellenistic times and later, a title could be attributed to a book based on its contents (cf Kirk and Raven (1960) 101), and terms like *Periodos Gēs* and *Periplus* became generic terms for a guide or travel book. See also on F9 and on F22.

2 He included Scythia (for him the area north of the Danube and the Black Sea, and the Caucasus, in book 1: *FGrH* 1 F184–194). Something about the east had become known by a voyage said to have been down the Indus and up the Arabian Gulf by one Scylax of Caryanda around 510–500. Book 2 had place and tribal names in Egypt, Ethiopia, Arabia, the Middle East, and India.

3 If Steph Byz sv Μασσαλία (Hecat *FGrH* 1 F58) quotes him verbatim, he treated Massalia as in Liguria, with Keltikē further inland.

a good deal of geography and ethnology; but his knowledge of the north was limited. He knew that the Istros (Danube, Latin Ister) flowed into the Black Sea, but also said that it rose in the land of the Celts near the 'city of Pyrene'.[4] He too had no idea that Britain (or Ireland) existed. Lurid tales about Ireland circulated in Strabo's time;[5] they are just the sort of thing Herodotos would have noted, if they had reached him. Parts of the Black Sea coast had been settled by Greeks, and they had tales about Scythia, the land to the north stretching into the unknown; he records details about alleged distant Scythian tribes which are more stories than fact (Hdt 4.8–36), including the Hyperboreans, further noted below.[6] As we will see in section 3, previous explorations past Gibraltar had sailed south, not north.

For areas beyond the Mediterranean and Black Sea, knowledge of distant parts came along old trade routes, and its quality can be judged by a passage in Herodotos. He records that gold comes from a sandy desert in India; it is dug up by ants larger than foxes. The locals get it at the hottest part of the year, as the heat makes the ants go underground more. Spices like frankincense and myrrh come from Arabia; the frankincense is from trees that are guarded by snakes, and can only be got by burning storax resin to drive away the snakes.[7] These were traders' tales; they had an interest in disguising the real origins of their goods, stressing the difficulties of getting them, and justifying their prices. In fact, gold and spices were shipped from the east to Yemen and Somalia, and then either transhipped up the Red Sea or carried overland through Arabia. Herodotos repeats them without comment, and then goes on 'I cannot speak accurately about the western edges of Europe, and I do not accept that there is a river flowing into the northern sea called Eridanos by the locals, from which it is said that amber comes; nor do I know of the Cassiterides ('tin islands'), from which tin comes to us'.[8]

A northern Eridanos is unique; it was usually identified with the Po in a myth about the origin of amber: see on F6. Perhaps Herodotos' source meant the northern Adriatic, and he misunderstood him. The 'tin islands' were an old traders' tale

---

4 Hdt 2.33.2; cf 4.49.1. It seems that Herodotos had no clear idea where the Celts lived, and either had misleading information about the source of the Danube, or misunderstood it. His 'Pyrene' suggests that he picked up something about trade routes over the Pyrenees, across France and southern Germany, and then along the Danube to the Black Sea: see Lloyd (1976) and Asheri *et al* (2007) *ad loc*. The actual source of the Danube is about 36 miles almost due east of Colmar.
5 Strabo 2.1.13: it is a cold wretched place; 2.5.8 and 4.5.4: the Irish are sexually promiscuous and incestuous savages living miserably in a cold climate, cannibals who eat their dead fathers; cf on F31.
6 If Greeks had folk memories of lands through which their distant ancestors had passed before settling in Greece, it is hard to discover them beyond place names associated with events in their myths (e.g. Phrygia).
7 Hdt 3.102–5, 106 (gold); 3.107–112 for extraordinary stories about the alleged origins of several spices.
8 Hdt 3.115. His reasons are that 'Eridanos' is a Greek word and so cannot be the name of a foreign river; and that he can find no one who has seen a sea 'on that part of Europe'; he adds that he accepts that tin and amber do come from a distant region. κασσίτερος, *kassiteros*, is the Greek for 'tin'. See also on F10.

on a par with those about gold and spices. Tin came from Cornwall (technically on an island), Brittany, and Spain, as noted above, p 6. Over the centuries, an original 'tin comes from a distant island in the north' (i.e. Cornwall) could easily become 'tin comes from distant islands in the north' in popular lore. It is curious that belief in this imaginary archipelago persisted even after the areas were under Roman control, and their actual origins were known; Ptolemy gave them a specific location in the Atlantic off north-west Spain.[9]

Hecataios and Herodotos show how little Pytheas, like all Greeks of his time, knew about the north. Map 1 shows how Pytheas probably envisaged the world.

He would have had a fair idea of the Mediterranean and probably the Black Sea, and could have drawn a recognisable map of that. He knew of the Mediterranean coast of Iberia, then Keltikē, the land of the Celts, France, with Massalia about halfway between Iberia and Italy. He will have known of the Pyrenees, but we cannot assume that he knew that they stretched all the way to the Atlantic, for him

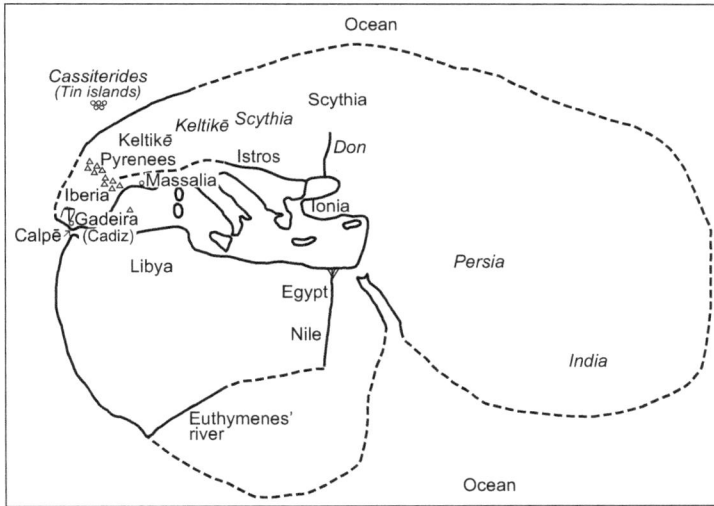

Map 1 Sketch map showing how Pytheas probably understood the world before departure. Ocean coasts are shown with dotted lines where he would believe they existed but have little or no knowledge of their actual shapes and directions. He would have heard of the places in italics, but have only vague ideas as to their location or extent. T = Tartessos.

Source: Drawn by the author.

9 So noted by Pliny, *NH* 4.119, 7.197; often mentioned without a number, ten is specified at Strabo 3.5.11. Ptol 2.6.73's location is 4° 45′ in his numeration. There are no islands anywhere in that region, except one or two rocks off the north coast of Spain. Attempts to identify them with, say, the Scillies or the Channel Islands are futile; tin is not found there.

part of the Ocean; nor can we assume that he knew how long it took to reach the Atlantic from the Mediterranean along their foothills, as noted on F34. Greece was across from Italy. Germany was unknown, and it must have been assumed that Keltikē extended eastward and became Scythia at some point; Greeks knew of Scythia from their Black Sea settlements. What lay further east was equally unknown. Across the water to the south, there was Libya, the general word for Africa west of Egypt, and then Egypt. He would know of its west coast from Euthymenes' voyage (p 12). Asia began in the east; in the Black Sea he may already have thought of the river Don as the dividing line (see on F30). Further east still, there were Persia and India, areas becoming better known from Alexander's conquests.[10] At the western end of the Mediterranean, there was Calpē, Gibraltar, past which you sailed into the Ocean: at some point the straits were known as the Pillars of Heracles.[11] Cadiz and Tartessos were a little way into the Ocean. The Ocean circled the world, because Homer had said so, but he had no idea of the actual contours of the Ocean coasts of Iberia and Keltikē; except that he had probably learnt from merchants coming from places on the Atlantic coast of Keltikē to Massalia how many days' journey they had variously come (see on F5 §4). If he had any idea of the overall shape of the Europe-Africa-Asia landmass, it would probably be some sort of ellipse, but with no idea of its real size. The tin islands apart, Britain, the Rhine, Jutland, and Scandinavia, and the north in general were totally unknown.[12]

Unknown, that is, unless we take into account the stories that circulated. Homer's Laestrygonians, who enjoy long daylight, have been noted on p 2, but they were poetic fantasy; Homer does not place them in the far north, but vaguely in the east,[13] and they are balanced with the Cimmerians, who live in perpetual night (*Od* 11.14–22).[14] If any of the tales of mysterious peoples in the Baltic such as

10 For Herodotos, India was the limit of the inhabited world: 3.98.2, 106.1–2.

11 Probably by Pytheas' time: see p 118.

12 For the Pyrenees, see on F34; for eventual knowledge of the Rhine, see the Note to Appendix 4 §1. Homer's Ocean is explicit at *Il* 18.607–8: Achilles' shield has the Ocean circling the world. After his introduction, Strabo begins his overview of the world by adopting Homer's geography, citing many passages to support his argument (1.1.3–10, esp 6–7). Crates of Mallos quoted *Il* 14.246, that the Ocean is the origin of everything, with an extra line that it flows all round the world (Plut *Mor* 938d). Strabo believed it: at 1.2.17 he says that Eratosthenes was wrong to doubt Homer. Herodotos, 2.21, 23; 4.8.2, 36.2, disbelieved the all-encompassing Ocean, though his reasoning is questionable (see Lloyd (1976, 1994) and Asheri *et al* (2007) *ad locc*).

13 At *Od* 12.4, Homer has Odysseus finally reaching a distant east, but his whole itinerary is poetic imagination, as pointed out by Heubeck and Hoekstra (1989) 48–59 on *Od* 10.80–132, and any attempt to trace it in the real world depends on one's own wishful thinking. Stories of long summer days and long winter nights in the far north could have been part of Greek folklore from their remote past, before they settled in Greece; or more recently from trade from the Baltic down Russian rivers, as suggested by Nansen (1911) 14–15.

14 Apollonius Paradoxographus, second century, in his *Historiae Mirabiles* (Giovanni 24 = Eudoxos of Rhodes *FGrH* 79 F2), says that there is a Celtic race that never sees the light. The story has been moved into the real world; but when that was published, the reality of the north was being established

Pliny could access centuries later in F17 already circulated, it merely emphasises how few firm facts were actually known. Several of our references touch on the Hyperboreans and the Rhipean mountains; (normally 'Ripean' in Latin). Each can be said both to exist and not to exist (though not, perhaps, in a Schrödinger sense). Hyperboreans ('Beyondnortherners' or 'Beyond the North Wind') was a generic name for any race or tribe believed to live somewhere beyond the Greeks' ken, vaguely north of the Black Sea.[15] They first appear in Hesiod's *Catalogue of Women* and the Homeric Hymn to Dionysios, c 700.[16] Herodotos has some delightful stories about them; curiously not from the Scythians, who he says do not know of them, though as their southern neighbours we might have expected otherwise, but from the island of Delos.[17] But no one had ever been there and met them.[18] As to the R(h)ipean mountains, ῥιπή, *rhipē*, means the force of a throw, and was used to refer to Boreas, the wind blowing from the cold north, from Homer onwards.[19] It was a small step to turn it into a proper noun for distant northern mountains where it was imagined that the wind originated.[20] Aristotle knew of tales that placed them in the arctic, but also as the source of the Danube and other large rivers.[21] In geographers from Eratosthenes onwards, the R(h)ipeans acquired a new lease of life, located somewhere in northern Europe or Scythia; in a given passage, it is often possible to suggest that they correspond to an actual range; for instance, they could correspond with the Carpathians in F17. But, just as no one had ever visited the Hyperboreans, no one had ever seen the R(h)ipeans. Strabo

by astronomers, and popular beliefs had to concede that northerners might have six months of day-light: see Appendix 2 §§5–10.

15 Nansen (1911) 15–22 collects in some detail accounts of the Hyperboreans and the trade routes from the north to the Mediterranean. Bridgman (2005) is an exhaustive study of the Hyperboreans; it includes a long section 3, chapters 5 to 9, discussing a handful of passages which either identified them with Celts or could be so read. So far as Pytheas is concerned, he claimed that Pytheas' names in F21, 'Gutones', Abalus, and Teutoni, were Celtic names (at 129); cf on 'Morimarusa', F17 n 8. But Pytheas would know that the locals there were not speaking a Celtic dialect. The most that can be said is that the few Celtic-Hyperborean references reflect general ignorance of the far north, or a vivid or poetic imagination.

16 Hes F150 M–W; cf Hdt 4.32; Hym Hom *Dion 29*; Pind *O* 3.16; *P* 10.30; *I* 6.23.

17 Hdt 4.32–6. Some of the stories connect them with Apollo and Delphi, as also in Bacchylides *Epinic* 3.58–62; Pindar *O* 3.16, with corroborative detail giving an air of verisimilitude to his narra-tive; and he then sensibly denies their existence by pointing out that if they were real, there should also be a race of 'beyond southerners' to the south.

18 Hdt 4.13 has a story of one Aristeas who travelled into the far north. If he did, he failed to leave any real details, and the whole tale Herodotos tells of the man, 4.13–17, verges on the unlikely.

19 Hom *Il* 15.171, 19.358. It is sometimes wrongly said that Homer spoke of the Ripean mountains, e.g. *OCD* sv *Rhipeae Monte*; he only uses *rhipē* for the cold north wind, Boreas. Hesiod says that a farmer in Boeotia should beware of Boreas blowing from Thrace in January and February (*Op* 504–8).

20 E.g. Alcman *PMG* 90; Hellan *FGrH* 4 F187b; Damastes *FGrH* 5 F1, who named three Hyperbo-rean tribes and said that it snowed continuously in the mountains.

21 *Meteor* 350b; cf 362b. He treats the stories as fanciful; it is arguable whether he disbelieved moun-tains in the far north as such or merely doubted that so many European rivers flowed from one range. As the source of the Danube, they took on a new lease of life with Apollonios Rhodios IV 282–6 ('whence the north wind blows'), the scholia on which noted that Timagetos *FGH* IV 519 had earlier said the same.

showed shrewd judgment in stating that neither existed: F39. When Pytheas set off, he can have had little tangible knowledge of what he might find.

There is a final point here. From c 600, Greeks had been going through Gibraltar to trade with Tartessos in southern Spain, and it is generally accepted that Marseille participated in this trade: see on F34. Would they not have picked up details about what lay further west, the Atlantic coasts of Iberia, for instance, from what they learnt in Tartessos? In turn, would not harbour talk in Marseille have filtered through to the rest of the Greek world? The question is complicated by the details of both Iberia and areas further north, including Britain, in the late poem by Avienus, c 350 AD, which had led to the suggestion that these details were first recorded at the time in a lost *Periplus* (n 1) by an unknown Massiliot. The short answers are that no such *Periplus* existed; Massiliots may have picked up something about the Atlantic and the north in the sixth century, but it would have fallen out of memory by Pytheas' time; if some of it was still recalled, no trace of it can be found in our Pytheas references. The extent of this trade is considered below, pp 14–16, and the Avienus aspect in Appendix 1.

3 Exploratory voyages before Pytheas

Pytheas did not have much by way of precedent. Massiliots were experienced mariners, but from trading within the Mediterranean and perhaps the Black Sea. Beyond Gibraltar, there had been trade with Tartessos, probably at the mouth of the Guadalquivir river, but this had probably diminished by his time ((pp 15–16), and see on F34). We know of only one Greek who had gone further and onwards into the Ocean: his fellow-countryman Euthymenes, also honoured with a statue in Marseille, as noted above, p 5. He is thought to have lived about a hundred years before Pytheas; he sailed down the west coast of Africa, until he reached a river which is usually identified with the Senegal, now the border between Senegal and Mauretania.[22] Other voyages were by non-Greeks. Pytheas would have known of two of them if he had a copy of Herodotos: a three year voyage by Phoenicians right round Africa from the Red Sea and back via Gibraltar in the time of the

---

22 A little further along the coast, there are the mouths of the Saloum, in Senegal itself, and then the Gambia within that country. Euthymenes' account, probably a *Periplus* (for the word see p 7 n 1), is mostly lost. He noticed that the Etesians caused the river to rise, and that it contained crocodiles and hippopotami. His suggestion that this was therefore the source of the Nile, which also periodically flooded and had these fauna, survives in several later writers. He cannot be dated more accurately than to the sixth or fifth centuries: we should probably think of a fairly early date, before Carthage controlled the trade with west Africa further south than Mauretania.

pharaoh Necho (610–595); and one by a Persian, Sataspes, in the time of Xerxes, c 500, who got close to the equator somewhere off Nigeria or Cameroun, until he hit the doldrums and turned back.[23] Two other early voyages are known, both by Carthaginians: Hanno and Himilco. Hanno sailed down the west coast of Africa; indeed, by c 450 Herodotos was able to record Carthaginian trade there (4.196). Both were later translated, Hanno certainly[24] and Himilco probably into Greek, and were known to Pliny. As to Himilco Pliny says that he was 'sent to discover the outside (extera) [coast] of Europe' (*NH* 2.169). How far he went depends on how we interpret the reference to him in Avienus, whose poem and its sources are discussed in Appendix 1. It speaks of him in lurid terms being becalmed in the Bay of Biscay with sea monsters (to vv 116–29). It comes just after the mention of Ireland and Britain noted on p 124, and the reference in vv 112–13 to Tartessos trading for tin in Brittany; but it is straining Avienus' text unduly to argue that Himilco included mentions either of Ireland and Britain, or the Tartessos' tin trade. Even if he did, Greeks would not know of it until he was translated, and we have no reason to think that that happened as early as Pytheas' time, and or that copies were then available in Massalia.

There is an intriguing mention in Pliny *NH* 7.197 of the otherwise unknown Midacritos. Within a long section naming those said to have been the first to have done or invented something, he says: 'plumbum <album> ex Cassiteride insula primus adportavit Midacritus': 'Midacritos was the first to bring tin from the island Cassiteris'.[25] The Cassiterides, as noted on p 8, were the 'tin islands', a non-existent archipelago believed to the source of tin; the singular 'Cassiteris' is only found here, and so could refer to Cornwall. However, Pliny's verb, 'adportavit', means 'brought'. While we can read it as meaning that he actually visited Britain and came home with tin, it could equally just mean 'imported'. In the absence of any other reference to him, we can only say that if he made an early voyage into the Atlantic, it was unknown to Hecataios of Miletos or Herodotos pp 7–8, and there is no reason to think that Pytheas knew about it either. Pytheas' voyage, as already noted, was essentially into unknown waters.

23 The round Africa trip, Hdt 4.42.2–4; Herodotos disbelieved the sailors correctly reporting that they had the sun 'on their right', i.e. to the north in the southern hemisphere. Sataspes' voyage is at Hdt 4.43.1–6.
24 The Greek translation of Hanno's voyage (*GGM* I 1–14) is said to have been taken from a dedicatory inscription in a temple. At its furthest point, he speaks of a lake of 'wild men and hairy women, whom our interpreters called gorillas' – the origin of our word.
25 The MSS just have 'plumbum', 'lead'; editors usually add 'album', 'white', the whole phrase meaning tin.

4 Relations with Carthage and passage through Gibraltar

Until Roseman (1994) 149–54, it was assumed that Pytheas used his own ship. She argued that the Carthaginians would not have allowed a Greek ship through the Straits of Gibraltar, and that Pytheas travelled from place to place on local trading ships, either as passenger or extra hand. She noted the evidence for trade along the coasts in question, adding that local captains would know the seas, and it met Polybios' sneer about Pytheas being poor, F30, because it would be cheaper. This idea was adopted by Cunliffe (2001) 56–8, with the variant that Pytheas sailed to the Massiliot colony of Agde, then overland up the Aude and down the Garonne to Bordeaux, and sailed north from there; at 58–60 he argues against accepting the 'Cadiz to Tanais' reference in F30, as too imprecise, and the 'Cadiz to Cape St Vincent' mention in F34 as possibly not from Pytheas. He overlooks the very precise meaning of *parōkeanitis*, the Ocean coast, in F30; while the context of F34, Iberia, and Pytheas' observation of the time taken to circumnavigate it also prove that he sailed round it. Be that as it may, the Carthage question needs noting.

(a) Carthage and Gibraltar

We have no actual text that says that Carthage closed the Straits to Greek merchantmen; it is an old inference which became dogma, based on the belief that Carthage destroyed the kingdom of Tartessos and took control of trade through the Straits; and treating fighting with Greeks in Sicily as Carthaginian aggression and total hostility to Greeks.[1] This is no longer tenable. Archaeology over the last 50 years or so paints a different picture; the area continued to flourish, and Greek goods continued to be imported. The 'aggression' is discussed in (b) below.

Southern Spain used to be a rich source of silver and tin. Greeks discovered it around 600 BC, and called it Tartessos; its geography is described on F34. Our earliest reference is the Stesichoros passage cited in F34, and by the poet Anacreon a generation later. Herodotos recorded a Phocaean tradition from that time that the *basileus* of Tartessos gave them enough money to build their city walls (1.163), and a Samian tradition that one Colaios traded there very profitably (4.152).[2] *Basileus*, conventionally translated 'king', was the usual Greek word for

---

1 So Dunbabin (1948) 434–4. 'That the straits were really blocked, through the fifth century and fourth . . . is all too familiar to need any documentation here': Hawkes (1977) 39–40. The traditional view of Carthage as the aggressor is encapsulated in the sub-heading 'The Carthaginian Invasion of Sicily', in Bury and Meiggs (1975) 189–90, relating to hostilities in 480.

2 Anacreon compared the kingdom's wealth to the cornucopeia, the horn of plenty: *PMG* 361, cited by Strabo 3.2.14, a page or so after F34. Herodotos 1.163 names the king 'Arganthonios', which would have overtones to a Greek of *argyros*, silver; though neither Anacreon nor Herodotos use the word 'silver'. The Phocaean version was that he gave them the money to build a wall round the city against a Persian invasion. There is a huge literature on Tartessos; see, for instance Celestino and López Ruiz (2016) 24–49; the archaeology and trade, including shipwrecks: Rouillard (2009), including map fig 5.1 at 132; Dietler and López Ruiz (2009) 299–312; Belén Deamos (2009); Aubet Semmler (2001b).

the chieftain of a foreign state (cf the comments on 'Basilia' on F6, pp 36–7); it does not follow that Tartessos had a hereditary monarchy. Neither kings nor chieftains usually give away money; the reality is that the trade was very profitable for the Greeks, who may have been dealing with the ancestors of the Turdetani (or perhaps the Turduli), the tribes who lived there, or at least their then chieftain (see on F34).[3] Bianchetti (1998) 27–8, 31–2 treats the presumed closure of the straits as an illusion, not Punic politics, and proposes that Massiliot traffic through Gibraltar never ceased, and merchants had continued access. Greek ceramic imports certainly continued. Shipwrecks contain Greek elements, thought they are less helpful on this point because they are in the Mediterranean, and there is no dispute about Greek presence on the Mediterranean coast of Spain.[4]

We may rule out a physical barrier at Gibraltar;[5] if we want to believe in the closure of the straits, an unfriendly reception at Cadiz would be sufficient to deter a hopeful Greek sea-captain. The clearest evidence that Carthage treated the western Mediterranean as their *mare nostrum* is her treaties with Rome made not long before Pytheas made his journey, the first c 348–7. Rome agreed not to trade west of a cape in Tunisia, Carthage making reciprocal concessions relating to Roman territory other than Rome itself. There is also a story in Strabo that a Roman ship trailed a Punic one to find out where tin was coming from, and the Punic captain ran his ship aground rather than risk capture, and was compensated for his loss by Cadiz.[6] On the other hand, Carthage herself needed friendly Greek ports in the west,[7] whether as the destination of goods or transhipment, or just to put in for food and water, and we may suspect that Marseille was one such.[8] A balanced view is that a Greek merchantman would not be prevented from passing Gibraltar and trading with Tartessos or Cadiz; whether over time Greek goods went more in Carthaginian and Punic ships is difficult to assess: the literary evidence is slender,

3 Or perhaps their predecessors, though archaeology does not suggest any violent displacement of one group of inhabitants by another.
4 Roseman (1994) 150 accepts that closure of Gibraltar cannot be proved. For continuity in Tartessos and Greek presence, see the authorities in n 2; also Olmost 1989; Mata (2001a), (2001b). Earlier discussions: Boardman (1999) 210–13, 215; Dunbabin (1948) 221, 233–5, 253–4.
5 I am obliged to Dr Roger Brock for drawing my attention to the texts in Ste-Croix (1972) Appx VIII, that two ships were needed to enforce a blockade successfully. That may well have been true to stop a trireme or penteconter, but the scourge of piracy shows that one oared ship could usually outmanoeuvre a merchantman which had to rely on the wind (cf p 5 n 2). But it is unreal to propose that Carthage or a Punic colony maintained a ship permanently off Gibraltar to stop the odd Greek ship coming through.
6 Treaties: Polyb 3.22–5 with Walbank (1976) *ad loc*; whether there were two treaties or three, and the dates of later one(s), are not relevant here. Three locations are proposed for the 'Fair Cape' (καλόν ἀκρωτηρίον, *kalon akroterion*): Cap Bon, in the north-east; Cap Farine, east of Bizerta; and Cap Blanc, north-west of Bizerta. Hammond (1986) 557–9 sees her treaties with Rome in the context of her attempt to gain control of Syracuse and the Straits of Messina in 345–37. Tin story: Strabo 3.5.10.
7 For the eastern Mediterranean, Carthage could use (say) Phoenician cities in Cyprus.
8 The continued Greek imports into southern Spain noted must either have been carried there on Greek ships, or loaded onto Carthaginian ships at friendly Greek ports.

but perhaps suggests that over time diminishing direct contact led to less accurate knowledge of the area. The Hermippos fragment, n 13, is perhaps neutral on this point. Herodotos 4.192.3, describing the animals of 'Libya', north Africa west of Egypt, includes γαλαῖ, *galai*, weasels or polecats, and says that they are 'similar to those in Tartessos'; this presupposes that odd details about Tartessos circulated in his time. A little later, Aristophanes *Frogs* 475 refers to the Tartessian μύραινα, *myraina*, the moray; it is a joke, with a play on words with 'Tartaros', the Underworld. It is not evidence for long-distance trade in fish with Athens.[9] Ephoros, in the fourth century, may possibly have referred to Tartessos and its metal trade, but in terms suggesting exaggerated stories rather than accurate knowledge.[10]

(b) Carthage and Greeks (mainly Sicily)

As to the alleged hostility, Phoenicia had founded colonies on the west coast of Sicily, as well as Carthage: Motye (Mozia), Panormos (Palermo), and Solous (Solu17to di Santa Flavia), important intermediate ports on her shipping routes.[11] When Phoenicia fell to the Babylonians in 573, Carthage took over the trading network, which extended down the west coast of Africa. By and large, relations with Greeks were amicable. Carthaginians were living in the Greek ports of Selinous and Syracuse in the fifth century, while in Motye, only Greek, not Punic, inscriptions have been found.[12] The reality is that Sicily was an island of several ethnic and cultural groups normally living and trading peaceably together: see *CAH²* IV 739–53 (Asheri). 'There is no reason whatsoever for dividing the island into sharply different, impermeable or antagonistic cultural areas' (*ib* 746).[13]

9 The *myraina* is the carnivorous moray ('murry' in Thompson (1947) 162); common in the Mediterranean. The god Dionysos and his slave want to enter the Underworld, and the door-keeper threatens to kill them and dispose of their bodies piecemeal: the *myraina* will eat their lungs. In the real world, no one would bring it alive from beyond Gibraltar to Athens; if it was exported, it would be salted. Dover (1993) 254 notes that it is not in Hermippos' list of imports, n 13. He cites later texts saying that it was a delicacy. The earlier is from Varro, first century; by his time, it could well have been a Roman import. If not, it would be no more than a guess from a superficial reading of Aristophanes. Note that the phrase was not copied from Euripides' *Theseus*: Kannicht (2004) 432, noting the confusion of the Aristophanic scholia.

10 Ps-Scymn 162–6, within Ephoros *FGrH* 70 F129b. The verses speak of tin flowing down a river from Keltikē, gold, and copper; but not silver.

11 If the Phoenicians had coastal settlements elsewhere on the island before Greeks came to Sicily, as stated in Thuc 6.2.6, they appear to have yielded to or amalgamated with Greek colonies with little or no resistance. Some, e.g. Dunbabin (1948) 22, regard it as improbable that they ever settled in the east.

12 Dunbabin (1948) 234–5; cf 252–4; 334. For Carthagian involvement in Sicily in 345–3, see n 15.

13 See the striking fragment of the comic poet Hermippos, late 5th century (F63 K-A; Gillula (2000)), which details the many goods imported into Athens from all over the Mediterranean world (incidentally showing how important was maritime trade generally). It includes dates and fine flour from Phoenicia and rugs and multi-coloured cushions from Carthage. Whatever discount one makes for comic exaggeration, it must basically reflect reality; and there is no reason to think that such commerce was limited to Athens and did not extend to other cities in the Greek world.

Where there was conflict, our sources are all Greek and present Carthage as the aggressor. But an objective reading paints a different picture. The 'aggression' was periodic, and was self-defence against Greek aggression: either against Greek attempts to found colonies in Punic areas,[14] or as a reaction to Syracusan aggression, for example to help neighbouring Greek cities in the west of Sicily against Syracusan expansion into their territory. Carthage saw that as equally a threat to the Punic settlements; her interest was to ensure that there were stable Greek cities in the west as a bulwark against Syracuse.[15] Outside these occasional incidents, relations were peaceful, as noted above. We should also note the same picture, resistance to Greek aggression, in the sea battle of Alalia.[16] Alalia in Corsica (now Aléria), like Marseille, had also been founded by Phocaea. Herodotos (1.166.1) says that they were always raiding local settlements. In reality, it was piracy, and in c 540 Carthage and the Etruscans joined forces to put an end to it. The Phocaeans abandoned Alalia, and Corsica, like Sardinia, fell into Carthage's sphere of influence.[17] In the absence of Greek aggression, Carthage had normal relations with the Greek world.

(c) The situation for Pytheas

Even on the outdated view that Gibraltar was closed to Greek merchants, it is unreal to propose a physical barrier of warships (n 5). Also, Pytheas was not a merchant; he only wanted to sail on into the Ocean. We can take this point further; we can infer that Marseille had a special relationship with Carthage. As noted above, she always needed friendly Greek ports. It is likely that Massalia was one of them, for instance for the transhipment of Spanish tin. Further, as noted in section 1, Marseille had trading settlements on the east coast of Spain in areas otherwise settled by Phoenicians. She was using these ports for trade with the

---

14 For detailed accounts, see Dunbabin (1948) 326–54; more briefly Bury and Meiggs (1975) 136, 186–7. The Greeks recorded it because it involved the deaths of two important Greek leaders, Pentathlos of Rhodes or Cnidos, killed at Lilybaion near Selinous, c 580, and Dorieus of Sparta, killed near Segesta, c 510.

15 The first time was in 480, and it is argued that it was done in liaison with Xerxes; so Bury and Meiggs (1975) 189–90 (cf n 1). Whether Carthage would ally herself to the power that had subjugated her Phoenician homeland is a question; the coincidence of dates may be no more than that; those asking Carthage for help could not know what Xerxes was planning or for which year. Hammond (1986) 269 takes a more balanced view. The second spanned 410–5, and the third various periods in the fourth century, and can only be assessed against the detailed background of Syracusan expansionist policies. Generally, Bury and Meiggs 385–90, 395–400; Hammond 471–3, 475–7. Hammond cites the authorities.

16 It is unclear if this is the same as a sea battle between Carthaginians and Phocaeans; as reported by Thucydides (1.13.6), it is said to have taken place when the latter were founding Massalia; the arguments are evenly balanced (Gomme (1945) 124–5; Hornblower (1991) 47).

17 At any rate, the Greeks abandoned Alalia and went to Elea (Velia, Castellamare di Velia), southern Italy. It is possible that some moved to Massalia, to explain there being two foundation myths for her (*CAH²* III.3 140–1). It is doubtful if Greeks ever settled in Sardinia; it was part of Carthage's bailiwick.

adjacent Spanish interior. It implies a *modus vivendi*; the settlements could not have happened without Carthaginian acquiescence. Hawkes (1977) 39–44 argues for an actual treaty between Carthage and Massalia in 326.[18] By that date, Rome's expansion had already caused Carthage to negotiate with Rome over trade routes, as noted above. Rome had also taken over Carthage's old ally Etruria, and that, and perhaps a perception that Alexander the Great might now expand westwards from Egypt, would encourage Carthage to look for new friends.[19] If the reality was closer to Bianchetti's view (see above), there would clearly have been no problem, as also proposed in McPhail (2014) 247–51.

(d) Cost

The other point taken by Roseman is cost. As a passenger, he would normally have been expected to pay for his journey. We must not be influenced by Polybios' comment; we do not know, except in general terms, the realities of paying for sea travel.[20] In any case, Polybios wanted to belittle Pytheas: see on F30. Pytheas almost certainly had practical support for his voyage, p 7; the real question there is how long his resources lasted, and therefore the length of his voyage, as to which see the Coda, pp 170–7. We may dismiss cost as an obstacle.

(e) Was a 'taxi service' a practical proposition?

We have adequate evidence of local trading as far as Britain, as Roseman summarises. But that does not help on the frequency of these voyages. It is unreal to imagine that Pytheas could arrive at one harbour and find another boat, loading up and ready to leave in a few days for the very next destination that he wanted. These trading voyages, for the most part, were probably limited to a few each

---

18 His wide-ranging argument includes Carthage protecting her tin trade from Galicia, the shifting balance of power as Rome expanded, and general concern that Alexander the Great was planning to conquer the west, noting the many embassies, including Carthaginian and Punic, that reached him around 324–3 (Diodoros 17.113.2; Arrian 7.23.2 for Greeks).

19 Rome's expansion had curtailed if not extinguished her long-standing alliance with Etruria. Since we can only date Pytheas' voyage approximately, the argument that Alexander's victories are relevant here is tenable, even allowing for the time for news of them to travel westwards; see pp 167–8 for using Alexander to date Pytheas.

20 Polyb 12.27.4 notes the expense of travel to see something for oneself as opposed to accepting what others have written about it; though his travels were subsidised by Scipio Aemilianus (Walbank (1972) 126–7); cf on F30 §7. Casson (1994) trawled through the literature and could find little direct evidence for actual costs. At 74 he assumes that payment would be with coins, by weight if not as currency, by money-changers at ports. He notes the costs of exit visas at Alexandria in 90 AD (154), the detailed costs of a land journey around 320 AD that has survived on papyrus (190–3), and a graffito from Puteoli, first century AD, listing the costs of a meal at an inn (207). Casson (1995) mentions passengers a number of times (see index), but only in the context of how they were accommodated.

year. Since Pytheas sailed north for several days before encountering the 'frozen sea' (Appendix 6 §§14–16), that must imply in his own ship. Then there is the question of how he reached Jutland and the Baltic. Roseman's analysis of trade does not include this area, and his ability to make that part of the journey is a further indication that he was in his own ship.

(f) Inferences from our Pytheas references

We must be cautious about an argument *ab silentio*, but there is no suggestion in our references that he was delayed at any stage while waiting for a ship to the next harbour; their tenor is that he was in his own ship. Further, in F30, Polybios does not attack Pytheas for being so poor that he had to use others' ships. Equally Strabo, for all his hostility, never implies that Pytheas was relying on others for his transport. We should particularly note F7: it may be significant that Pytheas is recorded as saying that the locals showed 'us' where the sun set: he was with his crew. With the precedent of Euthymenes, p 12, it is hard to see Pytheas thinking of exploring unknown lands other than in his own ship.

5 The astronomical background

Pytheas' education must have included a significant element of astronomy, and it is reasonable to infer that it included the latest thinking, that of Eudoxos of Cnidos, 391/0–c 337, a generation before Pytheas. Some propose that he was an actual pupil of Eudoxos; but it is just as consistent with having a good tutor familiar with Eudoxos' works.[1] Eudoxos was 'largely responsible for turning astronomy into a mathematical science'.[2] He attended Plato's lectures, and also studied with Archytas of Taras (Taranto), a Pythagorean mathematician then in Athens.[3] His *On Speeds*, noted on F8, reflects Archytas' influence but also underlines his own mathematical skills. Other works included *Phaenomena*, '*Things Seen*', which described not only the celestial sphere and the constellations, but also proposed various markings of it: the (celestial) equator, tropics, arctic circle, as well as

---

1 By the first century, Marseille had 'university' status: Cic *Pro Flacc* 63; Strabo 4.1.5; Tac *Ag* 4.2; Roller (2018) 178 also notes Pliny *NH* 29 (not 19) 10, 22–3, and suggests that this included a medical school. Ogilvie and Richmond (1967) 143, commenting on the Tacitus *Agricola* reference (see on F19), also note the reference to 'Massiliot customs' in Plaut *Cas* 963, set in Athens and adapted from a play by the Greek comic playwright Diphilos, Pytheas' contemporary. It is tempting to think that there was already an intellectual side to Massiliot life when Pytheas was a teenager. Müllenhof (1870) 324 said that it was 'nicht unwahrscheinlich', 'not unlikely' that Pytheas had been Eudoxos' pupil; Bianchetti (1998) 37–8 merely noted this as a possible connection between them.
2 Goldstein and Bowen (1983) 332.
3 An additional point is that in an institution such as the Academy in Athens, a student would absorb ideas that circulated orally but were never recorded in writing.

meridians and colures, by reference to the particular stars which lay on them; there was a second edition, *Enoptron*, '*Mirror*'.[4] He also pioneered the construction of sundials (Appendix 3).

As to terrestrial astronomy, by the end of the fifth century 'no Greek writer of any repute . . . conceived of the earth as anything other than a globe', stationary in the middle of the celestial sphere (Dicks (1970) 72.[5] One consequence was that lines such as equator and tropics, originally placed on the celestial sphere, could also be applied to the spherical earth. It was then 'only a short step to the realization that the length of the longest (or shortest) day defined the position of the equator and the tropics vis-à-vis a particular horizon, and that this in turn determined the place of the observer on the earth's circumference. Certainly Eudoxos had reached this stage of understanding' (Dicks (1970) 23, 154–5).

It is probable that we can glean some of Eudoxos' thinking about terrestrial latitudes from Aristotle, who at *Meteor* 362a 32–b 9 says that the earth is divided horizontally into five 'sectors' (τμήματα, *tmētata*), a mathematical term for the segments of a circle: cold uninhabitable zones in the far north and far south, a hot uninhabitable zone around the equator, and two temperate inhabitable zones between them.[6] It is very probable that his source, or one of them, was Eudoxos' horizontal divisions of the earth. Eudoxos would also have been in his mind at *De Caelo* 298a 16–18, that mathematicians have calculated the earth's circumference as 400,000 stades (45,455 miles, 72,730 km).[7]

Eudoxos is recorded as taking *gnomon* readings (Appendix 2 §3), enabling him to state the ratio of day to night at the summer solstice: 12:7 'for Greece' in his *Phaenomena*, and 5:3 in the *Mirror* (Hipparchos' *In Arat* (as in F8) at 1.3.9–10 and 1.2.22 (Lasserre (1966a) F67, F68)). It is uncertain whether the different readings reflect the limits of accuracy, or taking a second reading for his second edition where he was then living. We can place a ratio of 12:7 at the summer solstice at 42° 21′ N, with 15¼ hours of daylight, and 5:3 at latitude

4 For the details of these celestial circles, see Dicks (1970) 155–8; Goldstein and Bowen (1983) 335; Harley and Woodward (1987) 140–3; Cautadella (2015) 117–21; cf p 28. Hipparchos' *Commentary* on Eudoxos, from which F8 comes, corrected a number of Eudoxos' constellation details.
5 Not universally: Leucippos and Democritos argued for a flat earth (*ibid* 80, 82). Probably a Pythagorean idea, Aristotle gave practical reasons to prove that it is a sphere. *De Caelo* 285b 25–30 argues that a sphere is the shape into which the weight of particles of matter will naturally coalesce, and 296b 24–298 b21 includes details such as the shape of the earth's shadow during lunar eclipses (297a 8–b 23, 297b 23–31). See Dicks (1970) 72–3, 197–8, noting also (n 379) Arist *Meteor* 365a 32. Dicks also considers that Plato thought of the earth as a sphere on the footing that parts of his Phaido only make sense on that basis: *ibid* 94 with n 126.
6 Which in turn led to Stoic beliefs as to habitable and uninhabitable zones: introduction to F24.
7 Cautadella (2015) 121 n 33; Dicks (1960) 76 raises the possibility that the circumference was Archytas' figure. If so, it would be a fair inference that he explored the implications of a spherical earth, and Eudoxos, as his pupil, was privy to his ideas. Eratosthenes' measurement is at p xx, with the modern ones in n 2.

41° N, with 15 hours. Both are acceptable figures for Athens and Cyzicos, where he spent much of his life.[8]

As noted in Appendix 2 §2, a *gnomon* ratio at midday on the summer solstice corresponded to the ratio of day to night at that place, but without trigonometry it could not be turned into a latitude.[9]

It is unfortunate that Eudoxos' thinking on latitudes is befogged by an academic debate as to his word for them. In his geography (Γῆς περίοδος, *Gēs Periodos*, 'Circuit of the World') he seems to have divided the earth into four parts, with a vertical meridian and at least one line of latitude, discussed at length in Cautadella (2015) 121–6; cf Lasserre (1966a) 240–1.[10] Strabo used this work for his description of Greece, saying that it comprises three peninsulas, the Peloponnese, Attica and Boeotia, and the remainder, with a εὐθεία, *eutheia*, 'straight line', separating the Peloponnese to the south from the rest of Greece to the north: 9.1.1–2; F350 Lasserre.[11] Strabo then adds that he (Eudoxos) was 'expert in σχήματα (*schēmata*) and κλίματα (*klimata*)'. *Schēma* means 'shape' or 'figure', hence the peninsulas. The basic meaning of *klima* was an inclination or slope, but by Strabo's time it had come to be used for latitude, either in our sense or for a strip of land in which it was considered that conditions did not vary much.[12] This clear statement has been doubted, because the phrase 'expert in *schēmata* and *klimata*' was used of Eratosthenes in the verse geography (*periēgesis*: see p 7 n 1) of Ps-Scymnos (c 100: vv 112–14), so it is argued that Strabo mistakenly applied it to Eudoxos. That

8 Athens, 37° 59′ N, has 14 hrs 48 mins daylight; Cyzicos, at 40° 23′ N, near Bandırma, Turkey, on the south coast of the Black Sea about halfway between Istanbul and Bursa), had 15 hrs 3 mins. See, generally, Dicks (1970) 154–5 and Lasserre (1966a) 260. He ascertained the day of the solstice by checking what constellations were visible: Aratos (for whom see on F8) vv 497–9. For several days either side of it, the changes in the length of daylight would be imperceptible to him; they need modern techniques to differentiate. But that detail does not affect his limits of accuracy. The 12:7 ratio is just a measurement, and did not mean a 19 hour day, as has been proposed: Bowen and Goldstein (1991) 234–5. Hours, as a unit of time, were yet to come: Appendix 3.
9 The formulae are in Dicks (1970) 19–23; for Hipparchos and trigonometry Heath (1981) II 252–9. It needs an angle for the sun's ecliptic, but that had probably been established by Oinipedes of Chios at 24° (Heath *op cit* I 138, 147, II 244; Dicks *op cit* 157).
10 As with Herodotos (p 7), he knew nothing of northern Europe. For the area north of Africa and west of Italy, only one fragment remains: F359 Lasserre, concerning Liguria (south-west France); see Gisinger (1921) 104.
11 The actual geography is wrong: Eudoxos' line ran from the Ceraunian mountains (Mt Çika and the Karaburun Peninsula, north of Himarë, Albania), to Cape Sounion, via the Crisean Gulf (the Gulf of Corinth), the Megaris, and all Attica. The latter stretch can be thought of as broadly west to east; but no mariner would think of the coast to the north of the Gulf of Corinth as on that alignment. But Eudoxos was a mathematician, not a mariner, and the description illustrates how, in a world without maps, a landmass could be visualised. We cannot say if Eudoxos extended this line further west or east.
12 Strabo 2.5.34 with Dicks (1960) 154–64. Hipparchos adopted Eratosthenes' 252,000 stades as the circumference of the earth; 1° of latitude was therefore 700 stades, 80 miles, 130 km, and conditions were said not to change much over 400 stades, 45 miles, 70 km, i.e. up to 200 stades north or south of a given latitude (cf Dicks 156).

assumes that (a) Strabo knew the poem, and while the identity of its author is not certain, none of Strabo's known sources is a candidate; (b) the use of the phrase in the poem prevented Strabo from using it for Eudoxos; alternatively (c) the phrase circulated so firmly attached to Eratosthenes that it could not be applied to another geographer; (d) Strabo, having copied from Eudoxos, then mistook him for Eratosthenes. The division of the Greek landmass into peninsulas speaks of Eudoxos, not Eratosthenes: Roller (2018) 496–7 suggests that Eudoxos thought of areas of land as geometric shapes.[13] The argument is a red herring; Eudoxos had a latitude, and probably called it a *eutheia*, the word used by Dicaiarchos of Messana (Messina) a generation later, as next noted.[14] Strabo then used the more modern *klima* to praise Eudoxos' expertise.

Dicaiarchos' latitude figures in some of our references, so it is convenient to note it here. He was more or less Pytheas' contemporary and a pupil of Aristotle.[15] Influenced by Eudoxos, he proposed a latitude from the Straits of Gibraltar through Sardinia, Sicily, the Peloponnese, Caria, Lycia, Pamphilia, Cilicia, the Tauros mountains, to Mount Imaos (theoretically the Himalayas, as *Imaos* reflects the local name, but in practice perhaps the seemingly endless mountains at the furthest reaches of Alexander's conquests).[16] It was a reasonable attempt, if not accurate by modern standards; it proved very influential: see on F27 and F31.

It is with access to that learning, and Eudoxos' thinking about celestial and terrestrial latitudes, that we can understand how Pytheas was able to place Thule as he did, on the arctic circle and summer tropic: see on F2. He would also be aware of the value of taking *gnomon* readings; the question whether he took more than that at Marseille (F25, F27, F31) is discussed in Appendix 2 §§3–6. He certainly enabled Hipparchos, some 200 years later, to place Marseille on the same latitude as Byzantium.

6 Pytheas sets off

If we gather the threads of the previous sections together, we can detect at least three motives for Pytheas' journey. One is the natural curiosity of an intelligent young man of good education, already skilled in navigation, to see what might lie beyond Gibraltar and in the Ocean; coupled, perhaps, with dissatisfaction at the stories and rumours that circulated, but also inspired by what his fellow Massiliot Euthymenes had done some generations earlier. Another is discovering the sources of tin and amber, with the possibility of being able to source them more

---

13 A few lines further on Strabo notes that Eudoxos described several of the shores as shaped, such as being concave.

14 Alternatively, perhaps, he might have used παραλλήλος, *parallēlos*, 'parallel', as in Eratosthenes, F25.

15 Some propose that he was born c 375; others a *floruit* of c 320–300, implying a birth date of c 360.

16 Agathem 1.5 = F110 W; Cautadella (2015) 123, 131. It is possible that he included a map (Dilke (1985) is uncertain at 30 but accepts it at 117); Bianchetti (2015) 131, and see on F27.

advantageously than at present. A third is his astronomical background, to see if conditions further north corresponded to what he had been taught was theoretically the case, and to record his findings for others with better mathematics and astronomical knowledge to use. When he set off, the north was unexplored, little known, and perhaps mysterious. He was embarking on a bold venture.[1]

1 For the suggestion that Alexander sent him to explore the north, see the Coda, p 167.

THE FRAGMENTS

F1 Aetios III 17.3, p 383 Diels

(a) Stobaeus *Anth [Ecl]* 1.38: Πῶς ἀμπώτιδες γίνονται καὶ πλήμμυραι . . . Πυθέας ὁ Μασσαλιώτης τῇ πληρώσει τῆς σελήνης καὶ τῇ μειώσει τὰς ἑκατέρου τούτων αἰτίας ἀνατίθησιν . . .

How low and high tides are caused . . . Pytheas of Marseille attributes the causes of these to the waxing and waning of the moon.

(b) Ps-Plut 897b: Πῶς ἀμπώτιδες γίνονται καὶ πλήμμυραι . . . Πυθέας ὁ Μασσαλιώτης τῇ πληρώσει τῆς σελήνης τὰς πλημήρρας γίνεσθαι τῇ δὲ μειώσει τὰς ἀμπώτιδας.

How low and high tides are caused . . . Pytheas of Marseille [says that] high tides happen with the waxing moon and low tides at the waning moon.

Aetios, late 1st–early 2nd century AD, was the author of an encyclopedia or compendium summarising the views of philosophers on a range of scientific topics, called *Collection of the Views of Natural Philosophers*, or similar.[1] The work is lost, but substantial citations survive. It is an invaluable source for many lost texts from Anaximander onwards,[2] with the limitations of being summaries (hence the slightly different wording here), and usually from intermediate authors and not the original. Any modern discussion of tides states that they are caused by the gravitational pull of the moon and, to a lesser extent, the sun, with the rotation of the earth a third factor; the latter concept obviously unknown to Greeks. Contrary to Roseman (1994) 103, the wording is not Pytheas' but Aetios' summary of the three who opted for the moon, and by observation: Pytheas, Posidonios, and Seleucos of Seleucia for the Red Sea.[3] This reference is also relevant for Pytheas'

---

1 The title is in slightly different terms in our two sources. This reference to tides is also found in Ps-Galen *Hist Phil* 88, attributed to Euthymenes (p 12), universally regarded as an error.

2 He started with Thales, but it is disputed whether Thales ever put his views in writing or they were first recorded by pupils and successors.

3 Posidonios' views on tides were much wider than Pytheas'. According to Aetios, he said that the moon was an indirect cause: it affected the winds, which in turn affected the tides. Given the weather

DOI: 10.4324/9781003181392-3

route: cf F34 and p 111. Aetios also notes Aristotle, Heraclides, and Dicaiarchos, who chose the sun, purely by inference;[4] and four who variously argued for natural movements of the sea and rivers.[5] Tides in the Mediterranean, with which Greeks were familiar, are modest; some call the Mediterranean tideless, though, as we see here, they had words for high and low tides. Pytheas could only have worked out the influence of the moon once in the Atlantic and saw that spring and neap tides corresponded to the phases of the moon. Posidonios seems to have based his observations at Cadiz (p xxi n 5). Pytheas' book mentioned several tides: at Cape St Vincent (F34), around Britain (F5 and F16), and in the Baltic (F21, Appendix 4 §8, and see on F13).

F2 Cleomedes I 4 197–231 Todd

[197–208] λέγεται γοῦν ἐν Βρεττανίᾳ περὶ Καρκίνον τοῦ ἡλίου γινομένου καὶ τὴν μεγίστην ἡμέραν ποιοῦντος ὡς ὀκτὼ καὶ δέκα ὡρῶν ἰσημερινῶν γίνεται ἡ ἡμέρα, ἐξ δὲ ἡ νύξ· ὅθεν καὶ φῶς εἶναι παρ' αὐτοῖς νυκτὸς κατὰ τὸν χρόνον τοῦτον, αὐτοῦ παρὰ τὸν ὁρίζοντα τοῦ ἡλίου παρατρέχοντος καὶ ἀποπέμποντος τὰς αὐγὰς ὑπὲρ γῆν· ὅπερ ἀμέλει καὶ παρ' ἡμῖν γίνεται, ὅταν πελάσῃ τῷ ὁρίζοντι, πολὺ τοῦ φωτὸς τὴν ἀνατολὴν αὐτοῦ προλαμβάνοντος· ὅθεν καὶ ἐν Βρεττανίᾳ νυκτὸς εἶναι φῶς, ὡς καὶ ἀναγινώσκειν δύνασθαι. καὶ γὰρ τοῦτό φασιν ἀναγκαιότατον εἶναι, παρὰ τὸν ὁρίζοντα τοῦ ἡλίου τότε τὴν πορείαν ποιουμένου καὶ οὐ διὰ τῶν βαθυτέρων τῆς γῆς ἰόντος, διὰ τὸ ἐλάχιστον εἶναι παρ' αὐτοῖς τμῆμα ὑπὸ γῆν τοῦ θερινοῦ κύκλου.

[209–231] περὶ δὲ τὴν Θούλην καλουμένην νῆσον, ἐν ᾗ γεγονέναι φασὶ Πυθέαν τὸν Μασσαλιώτην φιλόσοφον, ὅλον τὸν θερινὸν ὑπὲρ γῆς εἶναι λόγος, αὐτὸν καὶ ἀρκτικὸν γινόμενον αὐτοῖς. παρὰ τούτοις ὁπόταν ἐν Καρκίνῳ ὁ ἥλιος ᾖ, μηνιαία γενήσεται ἡ ἡμέρα, εἴ γε καὶ τὰ μέρη πάντα τοῦ Καρκίνου ἀειφανῆ ἐστι παρ' αὐτοῖς, εἰ δὲ μή, ἐφ' ὅσον ἐν τοῖς ἀειφανέσιν αὐτοῦ ὁ ἥλιός ἐστιν. ἀπὸ δὲ ταύτης τῆς νήσου προϊοῦσιν ὡς ἐπὶ τὰ ἀρκτικά, ἐκ τοῦ πρὸς λόγον καὶ ἕτερα μέρη πρὸς τῷ Καρκίνῳ γίνοιτ' ἂν ἀειφανῆ τοῦ ζῳδιακοῦ. καὶ οὕτως, ἐφ' ὅσον τὰ παρ' ἑκάστοις φαινόμενα ὑπὲρ γῆς αὐτοῦ διέρχεται ὁ ἥλιος, ἡμέρα γενήσεται. καὶ ἔστι κλίματα τῆς γῆς ἀναγκαίως, ἐν οἷς καὶ διμηνιαία καὶ τριμηνιαία γίνεται ἡ ἡμέρα, καὶ τεσσάρων καὶ πέντε μηνῶν. ὑπὸ δὲ τὸν πόλον αὐτὸν ἓξ ζῳδίων ὑπὲρ γῆς ὄντων, ἐφ' ὅσον ταῦτα διέρχεται ὁ ἥλιος ἀειφανῆ ὄντα, ἡμέρα γενήσεται, τοῦ αὐτοῦ κύκλου καὶ ὁρίζοντος

---

that can be encountered in the Atlantic, this was a reasonable inference; but he said rather more about tides than that: see F214–F220 E–K; commentary Kidd (1988) 759–92.

4 The argument appears to have been that as the sun gets hotter towards midday, it has a greater effect on the sea: cf Arist *Meteor* 366a 19.

5 As armchair deduction with the Greeks' mindset, it is hard to improve on Socrates in Plato's *Phaido* 114c–112a, that tides were caused by the waters flowing back and forth into an underground chasm which Homer had described (*Il* 8.14), and later poets called Tartaros. Homer's geography was widely believed: cf on F35. In addition to Plato, Aetios cites Timaios, Crates, and Apollodoros.

καὶ ἀρκτικοῦ γινομένου αὐτοῖς καὶ ἰσημερινοῦ. τοῖς μὲν γὰρ ἐν Θούλῃ συμπίπτει ὁ θερινὸς τροπικὸς τῷ ἀρκτικῷ· τοῖς δ' ἔτι ἐνδοτέρω ὑπερβαίνει ὁ ἀρκτικὸς τὸν θερινὸν εἰς τὰ πρὸς τὸν ἰσημερινὸν μέρη, ἐκ τοῦ πρὸς λόγον τούτου γινομένου· τοῖς δ' ὑπ' αὐτῷ τῷ πόλῳ ὁ ἰσημερινὸς τὰς τρεῖς λαμβάνει σχέσεις, ἀρκτικὸς μὲν γινόμενος, ὅτι περιλαμβάνει τὰ ἀειφανῆ τῶν ἄστρων μηδενὸς ἁπλῶς παρὰ ταύτοις ἢ δυομένου ἢ ἀνίσχοντος, ὁρίζων δὲ γίνεται, ὅτι χωρίζει τὸ ὑπὲρ γῆς τοῦ κόσμου ἡμισφαίριον ἀπὸ τοῦ ὑπὸ γῆν, ἰσημερινὸς δέ, ὅτι αὐτός ἐστιν ὁ διαιρῶν αὐτοῖς εἰς ἴσα τὴν ἡμέραν καὶ τὴν νύκτα, ὃς καὶ τοῖς ἄλλοις μὲν πᾶσιν ἐπ' ἴσης ἰσημερινός ἐστιν, οὐκέτι δὲ οὔτε ὁρίζων οὔτε ἀρκτικός . . .

[197–208] *So it is said that in Britain, when the sun is in Cancer and making the longest day, the day is about 18 equinoctial hours, and the night 6 hours; consequently there is light there at night during this time, since the sun runs along the horizon and sends its rays above the earth. This indeed also happens with us: when the sun approaches the horizon, much of its light anticipates its rising. So in Britain there is light at night, so that it is possible even to read. They say that this must necessarily be the case, because the sun makes its journey along the horizon and does not go through the deeper parts of the earth, because there the segment of the summer circle below the earth is very small.*

[209–231] *Around the island called Thule, where they say the philosopher Pytheas of Marseille was, it is reported that the whole of the summer [tropic] is above the earth, also becoming the arctic [circle] there. In this place, when the sun is in Cancer, the day will be a month long, at least if all parts of the Cancer are always visible to them; if not, for as long as the sun is in the parts of it [Cancer] that are always visible. As one goes further north from this island, by the same reasoning other parts of the zodiac will always be visible apart from Cancer. Therefore to the extent that the sun passes through what is visible [sc of the zodiac] above the earth in different places, there will be daylight. It necessarily follows that there are latitudes of the earth where the day lasts for two or three months, and four and five months. Since under the pole six signs of the zodiac are visible, as long as the sun goes through these ever visible signs, it will be daylight, since the same circle is the horizon, the arctic circle, and the equinoctial circle in this place. For those on Thule the summer [tropic] coincides with the arctic [circle]. For those further north, the arctic [circle] exceeds the summer [tropic] in proportion to the parts [of the heaven] approaching the equinoctial circle, in the same proportion. For those under the pole, the equinoctial [circle] takes on the three aspects: it is the arctic [circle], because it includes the stars which are always visible, with none of them ever either setting or rising; it is the horizon, because it separates the hemisphere above the earth from the one below it; and it is the equinoctial [circle], because it divides the day and night into equal parts; that is true elsewhere, but it is no longer either the horizon or the arctic [circle]. . . .*

26

Cleomedes, probable date c 200 AD, was a Stoic philosopher whose surviving work is a manual of astronomy called *The Heavens* or similar, thought to be based on a series of his lectures.[1] It contains no original research, but is an important source for earlier thinking, including Posidonios (p xxi).[2] Cleomedes has just said that the lengths of day and night differ at different latitudes because the sun passes through different parts of the zodiac circle; since this is at an angle to the observer's horizon, some part of the zodiac is always below the earth at any given latitude.[3]

[197–208]: For the term 'equinoctial hours', which post-dates Pytheas, see on F7. Hipparchos, with more data, and knowledge of trigonometry, had improved and expanded Eratosthenes' figures for daylight lengths (Appendix 2 §§5–7; cf on F28); but Cleomedes and Pliny F15 show that other lists existed with slightly different figures. Cleomedes' 6 hours of night in Britain at the summer solstice is repeated at II 1 438–44, where he says that ἱστορεῖται, *historeitai*, 'it is reported' what the figures are for eight places from Meroë northwards. Expressed as the corresponding daylight figures, the list would read

Meroë 13 hrs
Alexandria 14 hrs
Hellespont 15 hrs
Rome over 15 hrs
Marseille 15½ hrs
Among the Celts 17 hrs
Britain 18 hrs

– which figures, especially the northern ones, may be compared to Hipparchos' figures noted on F28. Pytheas knew about long summer days, but had no means of measuring them: Appendix 3; cf Appendix 2 §9. Eighteen hours is in fact true for the latitude of the Cairngorms and the Inverness-Aberdeen area, 57°–58°; Hipparchos' table has an entry which corresponds to 58° 13' N.[4] That at midsummer the sun 'runs along the horizon' goes back to Pytheas' account of short nights: see F7. There is a useful discussion in Roseman (1994) 104–9 as to how Cleomedes himself envisaged all this.

---

1 The title of the translation by Bowen and Todd (2004) (available online) is 'Cleomedes' lectures on Astronomy' with the subtitle 'The Heavens'; see *op cit* 1 n 1. For his date, and Cleomedes as a teacher, 2.
2 In a nutshell, for the structure of the universe and the relationship between physics and astronomy: Bowen and Todd (2004) 5–7, 15–17.
3 I 4 184–96. Cleomedes' description of the zodiacal circle: I.4.30–43; Bowen and Todd (2004) 51–2 with Fig 6 on p 174.
4 Dicks (1960) 193. If Cleomedes' dates were later rather than earlier, there is a slight possibility that he could also have accessed figures for Britain based on daylight measured by Agricola's staff on his final expedition, 84 AD; wherever we locate Mons Graupius, the most northerly Roman camp is at Auchinhove, about on the latitude of Inverness (Tac *Ag* 29–38; Ogilvie and Richmond (1967) map 5 p 61; 64–5; 251–2).

[209–231]: Cleomedes now speaks of the various lengths of daylight in the far north. His reference to Pytheas is second-hand ('they say' that he was at Thule), but F31 confirms that it was Pytheas who located Thule in relation to the arctic circle and summer tropic. Were these celestial or terrestrial lines? The summer tropic (*tropē*, 'turn') was a slanting line on the celestial sphere representing the sun's path through the heavens; at midsummer, the sun had reached its highest point and then 'turned'.[5] He knew that there was continuous daylight on Thule: see Appendix 2 §8, so the sun had reached its highest point there. The celestial arctic circle, however, was not a fixed line; it depended on one's latitude. 'To the Greeks [it] meant a circle on the celestial sphere, parallel to the [celestial] equator and tangential to the observer's horizon, which marked the limit of the circumpolar stars that were always visible and that did not set below the horizon at that particular latitude. . . . Hence every latitude has a different 'arctic circle'.'[6] It is probable that Pytheas' arctic circle was also celestial, though the contrary is argued.[7] However Pytheas expressed it, Eratosthenes could use it to see that Thule in the north corresponded to Syene in the south, and was on the terrestrial arctic circle on our sense: see his calculation set out on F25.

We cannot, however, attribute the month of daylight to him, as some have (e.g. Tierney and Bieler (1967) 115). The point is considered in more detail in Appendix 2 §8; briefly (a) if Pytheas visited Thule, it was not in high summer; (b) he lacked both data and the mathematics to calculate it; (c) his knowledge depended on what he understood his Shetland informants told him, which was their own belief on the point. Cleomedes was not following the calculations of Geminos, 6.6–42, also noted in the Appendix, which would have given him a 24 hour day at the summer solstice; he is encapsulating some theorising based on the assumption that parts of the zodiac, including some or all of Cancer, would be visible (see Roseman (1994) 105), except, presumably, when there is continuous daylight. But his calculation is also likely to draw on Eratosthenes using Pytheas to put Thule on the terrestrial arctic circle, as just noted. This is the only reference which could

5 Dicks (1970) 155, 158, citing Hipparchos *Comm In Arat* 1.2.18, 1.10.13–15 (summer); 1.2.20, 1.10.17 (winter).
6 Dicks (1960) 165–6; cf *id* (1970) 31, 155; Cleomedes explains this at length, e.g. I.1.195–208, I.3.22–43. The 'arctic circle' was so called because the Great Bear never sets at the latitude of Athens, and ἄρκτος, *arktos*, means 'bear'. Note, however, Kidd (1988) 744, as to whether Posidonios proposed a fixed arctic crcle in our sense; it would imply building on his knowledge of Pytheas. Bowen and Todd (2004) fig 10a p 179 reflects the two (celestial) lines coinciding for Thule.
7 'Pytheas put Thule at 66°': so Kidd (1988) 745 on Posid F206 E–K ≈ F33. The concept of a terrestrial arctic circle was part of astronomical thinking in Pytheas' time (Arist Meteor 362a 32–b 9 cited on p 20), but if Kidd is basically right Pytheas cannot have said 66°. When Eratosthenes calculated the circumference of the earth he divided it into 60 parts (pp xx–xxi with Strabo 2.5.7); it was Hipparchos who first divided a circle into 360° (Heath (1921) II 216). We must not retroject the clear exposition of celestial and terrestrial circles centuries later in Cleomedes I 193–208 to Pytheas. Geminos (see on F7) was to deal with the celestial arctic circle in his Chapter 5, but does not mention a terrestrial circle elsewhere.

possibly suggest that Pytheas recorded constellations, apart from that to the pole star in F8.[8] If he noted what parts of which constellations were more (or less) visible as he moved north, it has not survived. Indeed, it is not clear that the use of constellations to determine latitude was understood in Pytheas' time; the concept of establishing latitudes was only just developing cf pp 20–1.[9]

F3 Cosmas Indicopleustes, *Christian Topography* 2.80

Πυθέας δὲ ὁ Μασσαλιώτης φησὶν ἐν τοῖς Περὶ Ὠκεανοῦ ὅτι παραγενομένῳ αὐτῷ ἐν τοῖς βορειοτάτοις τόποις ἐδείκνυον οἱ αὐτόθι βάρβαροι τὴν ἡλίου κοίτην, ὡς ἐκεῖ τῶν νυκτῶν ἀεὶ γενομένου παρ' αὐτοῖς.

Pytheas of Marseille says in his On the Ocean *that when he arrived in the most northerly places the locals showed him the sun's bed, as there the nights are continuous for them.*

Cosmas was an 'Alexandrian merchant, Nestorian,[1] and argumentative autodidact' (so the *OCD*) of c 550 AD, whose *Christian Topography* argued that a spherical earth was a pagan idea: the world was shaped like Moses' tabernacle. On the bottom was the flat earth, with a domed sky above. He argued that the sun disappeared each night behind a cone-shaped mountain at the edge of the world; whether behind its base or higher up determined the different lengths of daylight. To 'prove' his case, he quoted snippets from classical authors. It is unclear if a copy of Pytheas' book survived in his time, or he cited him from an intermediate author. What Pytheas said about continuous days and nights is considered in Appendix 2 §8. The 'sun's bed' is presumably from the same part of Pytheas' book from which Geminos took F7, where it is noted. Cosmas, apart from his intrinsic interest, preserves the title of Pytheas' work, also in F7.[2] There are helpful discussions of Cosmas' views in Thomson (1948) 387–8 and Roseman (1994) 144–5; see also Windstedt (1909) 6–9 (an 'indescribable medley'). It was always a minority Christian view.[3]

---

8 It was based on theory, not only because no one had been there to make observations, but also because you cannot know which stars rise or set during periods of perpetual daylight. See also Appendix 2 §8.

9 '[A] way to determine latitude was the relative height of celestial bodies other than the sun, something that became apparent when Greeks began to venture regularly outside the narrow latitudinal limits of the Mediterranean': Roller (2010) 170–1, who then notes this reference and also F2 and F31 for the references to the arctic circle. He assumes that Pytheas also noted which constellations did not set at that latitude. However, Roller's comment seems more relevant to Greek penetration into Egypt and India after Alexander's conquests.

1 Nestorius, Archbishop of Constantinople from 438–441, taught that the divine and human aspects of Christ were distinct, a heretical notion particularly in western Christianity.

2 Note that Cosmas gives the title as *Peri Oceanou*, without the *tou* ('the') of F7; it is usual to cite it with the *tou*.

3 Roseman notes the suggestion that Cosmas saw the world as God's footstool. See also the reference to Bishop Severian and Basil of Caesaria in en.wikipedia.org/wiki/Flat_Earth#Early_ Christian_Church. Basil hedged his bets on the point.

F4 Diodoros Siculus 5.21.3–4 ≈ Timaios *FGrH* 566 F164.21

[3] αὕτη γὰρ τῷ σχήματι τρίγωνος οὖσα παραπλησίως τῇ Σικελίᾳ τὰς πλευρὰς οὐκ ἰσοκώλους ἔχει. παρεκτεινούσης δ' αὐτῆς παρὰ τὴν Εὐρώπην λοξῆς, τὸ μὲν ἐλάχιστον ἀπὸ τῆς ἠπείρου διεστηκὸς ἀκρωτήριον, ὃ καλοῦσι Κάντιον, φασὶν ἀπέχειν ἀπὸ τῆς γῆς σταδίους ὡς ἑκατόν, καθ' ὃν τόπον ἡ θάλαττα ποιεῖται τὸν ἔκρουν, τὸ δ' ἕτερον ἀκρωτήριον τὸ καλούμενον Βελέριον ἀπέχειν λέγεται τῆς ἠπείρου πλοῦν ἡμερῶν τεττάρων, τὸ δ' ὑπολειπόμενον ἀνήκειν μὲν ἱστοροῦσιν εἰς τὸ πέλαγος, ὀνομάζεσθαι δ' Ὅρκαν. [4] τῶν δὲ πλευρῶν τὴν μὲν ἐλαχίστην εἶναι σταδίων ἑπτακισχιλίων πεντακοσίων, παρήκουσαν παρὰ τὴν Εὐρώπην, τὴν δὲ δευτέραν τὴν ἀπὸ τοῦ πορθμοῦ πρὸς τὴν κορυφὴν ἀνήκουσαν σταδίων μυρίων πεντακισχιλίων, τὴν δὲ λοιπὴν σταδίων δισμυρίων, ὥστε τὴν πᾶσαν εἶναι τῆς νήσου περιφορὰν σταδίων τετρακισμυρίων δισχιλίων πεντακοσίων.

[3] *It [Britain] is triangular in shape, very much like Sicily, but its sides are not equal. It stretches obliquely along the coast of Europe. The headland which is nearest the mainland, which is called Kantion, they say is about 100 stades [11 miles, 18 km] from the land, at the place where the sea has its outlet; the second headland, called Belerion, is said to be four days voyage from the mainland; and the last they report extending into the [open] sea is called Orca. [4] Of the sides the shortest, 7,500 stades [850 miles, 1,360 km], lies alongside Europe; the second, from the crossing to the [northern] tip, is 15,000 stades [1,700 miles, 2,720 km]; and the other is 20,000 stades [2,270 miles, 3,630 km], so that the entire circuit of the island amounts to 42,500 stades [4,830 miles, 7,730 km].*

Book 5 of Diodoros' *Universal History* is basically an epitome of Timaios: p xxi. Pytheas had just brought the existence of Britain, and that it was triangular, to the notice of the Mediterranean world. As a native of triangular Sicily, Timaios would be interested; as in F5 and perhaps F6, we are reading Pytheas at second hand.[1] Anything else about Britain would be texts retailing what circulated orally from traders; no other author visited Britain until Caesar.

§3 Diodoros has just said that there are islands in the Ocean, Britain being the largest. The measurements must be from Pytheas. As to its comparison with Sicily, on a modern map it appears trapezoid, but Greeks and Romans thought of it as triangular. Thucydides says that 'trinacria', 'with three headlands', was its old name.[2] It is unclear if the comparison of Britain with Sicily was Pytheas' or

---

1 So Walbank (1956) 394 on Polyb 3.57.3, referring to F5: 'Pytheas' account of Cornish tin mines survives via Timaeus'; Mette (1952) 14 n 2, on Diodoros 5.21.5, printed below ('durch Vermittlung des Timaios . . . auf Pytheas').
2 Thuc 6.2.2. The three promontories were named as Capo Lilibeo in the north-west, Capo Peloro in the north-east, just across from the Italian mainland, and Capo Passero in the south-east (Lilibaeum, Pelorum or Peloris and Pachynum: Polyb 1.42.3–7; Plin *NH* 3.87). As *trinacria* is pure Greek, it cannot reflect pre-Greek inhabitants such as the Sicels, but it is intriguing that Homer has Odysseus visiting an island Θρινακίη, *Thrinakriē*, (*Od* 11.107, 12.127, 12.261), suggesting very old trading contacts:

Timaios': either is certainly arguable; but its shape if not its old name would be known to mariners in Marseille, and we can support the comparison being Pytheas' if we think that the Sicily-shaped island in Hecataios of Abdera's *On the Hyperboreans* was inspired by Pytheas, as to which see p 167. Mela, F12, noting the comparison, is neutral on this point. Two details in Diodoros are probably close to Pytheas' actual words. The word translated 'obliquely', λοξῆς, *loxēs*, 'slanting', is probably how Pytheas described how the south coast of Britain splayed away from the French coast as he sailed west along the English Channel, as referred to below. Where Kent is close to the mainland, Pytheas perceived the English Channel as a funnel, from which the sea flows into the North Sea; hence ἔκρουν, *ekroun*, 'a flowing out' in §3 and πορθμοῦ, *porthmou*, 'passage' in §4 (literally, both ferry and narrow waterway). It is so shown on map 4b, p 179. But some later geographers seem to have understood the word to mean, or include, Pytheas' 15,000 stades coast, the east coast, so that that also faced continental Europe, to judge from Mela's description, F12 (map 2 p 51).

Secondly, the names come from Pytheas: Britain for the whole island, for which see on F5, and the three British headlands: Belerion (Land's End), Orca (the John o'Groats area, probably Dunnet Head or Duncansby Head), and Kantion (Kent; the Dover area: perhaps he meant South Foreland). Headlands were essential in ancient navigation; cf on F9 n 2. He also named various offshore islands: see on F19.

The Straits of Dover figure is probably from Pytheas. While it looks wrong to us, it is a reasonable guess if we envisage him sailing home off the French coast and seeing Britain on his starboard side.[3] The four days from Land's End to the mainland cannot refer to a direct sail across to the French coast opposite. From Porthcurro, for example, the nearest beach to Land's End, to the French coast in the vicinity of Morlaix or Lannion is around 120 miles, 190 km, perhaps a day and a half's overnight sail; by whatever stages Pytheas reached Cornwall from Brittany, he would know that the final stretch across the Channel was of that order. His four days makes sense on the basis that he learnt it at Corbilo on the Loire estuary (see on F36 and the Coda, p 172). It was probably one of the places to which Cornish tin was shipped: see on F5. That is some 350–375 miles, 560 to 600 km, and four days and nights is realistic for those familiar with the route and able safely to negotiate the several large bays on the Breton coast, and avoid the dangerous stretch around Pointe du Raz (see on F26); see pp xvii–xviii for sailing times and the Coda p 167.

Greek settlement in Sicily began around 700 (*CAH²* 94–113 seqq, 163–86 (Graham)). ἄκρα, *akrē*, literally a point, also, like ἀκρωτήριον, *akrōtērion*, in the text, meant a headland or promontory. *Thrinax* itself meant a trident or a three-pronged (agricultural) fork. See also Hornblower (2008) 268; Gomme (1970) 211.

3 If he accessed local information, his informants could not have known either 'hours' or 'stades', but could express it in general terms, e.g. 'part of a day'. Two hundred stades (about 23 miles, 36 km) would be accurate, but an error in transmission here is unlikely. For this part of his journey home see the Coda, pp 176–7.

§4 The reference to the 'passage', Dover to Calais, shows that Pytheas' 15,000-stade figure is the east coast and the 20,000 stades the west coast, Land's End to John o'Groats. See above for Mela's view; Strabo's description is in Appendix 7.[4] All the figures are too high by a factor of around 10%, as discussed in detail in the Coda, §5 paras 5 8 and 11 (pp 173–4, 175, 176–7), where it is assumed that he calculated the length of the south coast on the assumption that it was more or less the same as his figure for the French coast opposite, along which he sailed when returning home. However we judge his three distances, he was the only person who had actually written from sailing along the west and east coasts, and traversing the English Channel; and his total of 42,500 stades became more or less standard: 'more than 40,000 stades' in Polybios, F30, and 4,875 (Roman) miles in Pliny, F18. They were very possibly also in Eratosthenes. Strabo did not offer a circumference, and variously gave figures just for the south coast of 4,300 or 4,400 stades (490, 500 miles; 785, 800 km) at 4.5.1, and 5,000 stades (570 miles, 910 km) at F25, though he treated this as its longest side, and his location of Britain in relation to Europe is difficult to assess, as noted in Appendix 7. Caesar had local information, probably expressed as so many days sailing, and turned it into Roman miles: west coast 800, east coast 700, south coast 500 (*BG* 5.13: 735, 650, 450 modern miles; 1,180, 1040, 720 km). Comparison with actual figures is difficult, partly because they depend on whether they are based on high or low tide marks, and to what extent estuaries and bays are included. At one extreme, the Ordnance Survey figure, which keeps close to the coast, offers 11,703 miles (18,725 km); at the other, if you just draw a parallelogram round the mainland the four sides total some 1,700 miles, 2,720 km.[5] Pytheas' total figure of some 4,830 miles, 7,730 km, happens to fall in the middle, and so even if somewhat overestimated is arguably reasonable.

Between this text and F5 there is a section describing Celtic Britain and its inhabitants (5.21.5–6). It is offered here in Oldfather's Loeb translation. It may be compared with what Strabo said, noted on F38, and how Pytheas described the British diet. There are hints of both, but it must be borne in mind, as noted on F38, that most knowledge of Britain reflected the south. Diodoros' contemporary Caesar said that he could not learn much before his first invasion, and was not much the wiser after his second (*BG* 4.20–1, 5.12–14);[6] that probably reflects traders' reticence: cf on F36.

[5] And Britain, we are told, is inhabited by tribes which are autochthonous and preserve in their ways of living the ancient ways of life. They use chariots, for instance, in their wars, even as tradition tells us the old Greek heroes did in the Trojan War, and their dwellings are humble, being built for the most part out of reeds or logs. The method they employ of

---

4 Ptolemy was to get England correctly oriented, but had Scotland leaning over to the right. That Britain slanted north-east is most likely how later geographers interpreted Pytheas' dimensions; cf on F12 and F27.

5 See the various options at brilliantmaps.com/gb-coastline/.

6 Even allowing for his primary interest being the natives' military strength.

harvesting their grain crops is to cut off no more than the heads and store them away in roofed granges, and then each day they pick out the ripened heads and grind them, getting in this way their food. [6] As for their habits, they are simple and far removed from the shrewdness and vice which characterize the man of our day. Their way of living is modest, since they are well clear of the luxury which is begotten of wealth. The island is also thickly populated, and its climate is extremely cold, as one would expect, since it actually lies under the Great Bear. It is held by many kings and potentates, who for the most part live at peace among themselves.

F5 Diodoros Siculus 5.22.1–4 ≈ Timaios *FGrH* 566 F164.22

[1] ἀλλὰ περὶ μὲν τῶν κατ' αὐτὴν νομίμων καὶ τῶν ἄλλων ἰδιωμάτων τὰ κατὰ μέρος ἀναγράψομεν ὅταν ἐπὶ τὴν Καίσαρος γενομένην στρατείαν εἰς Πρεττανίαν παραγενηθῶμεν, νῦν δὲ περὶ τοῦ κατ' αὐτὴν φυομένου καττιτέρου διέξιμεν. τῆς γὰρ Πρεττανικῆς κατὰ τὸ ἀκρωτήριον τὸ καλούμενον Βελέριον οἱ κατοικοῦντες φιλόξενοί τε διαφερόντως εἰσὶ καὶ διὰ τὴν τῶν ξένων ἐμπόρων ἐπιμιξίαν ἐξημερωμένοι τὰς ἀγωγάς. οὗτοι τὸν καττίτερον κατασκευάζουσι φιλοτέχνως ἐργαζόμενοι τὴν φέρουσαν αὐτὸν γῆν. [2] αὕτη δὲ πετρώδης οὖσα διαφυὰς ἔχει γεώδεις, ἐν αἷς τὸν πόρον κατεργαζόμενοι καὶ τήξαντες καθαίρουσιν. ἀποτυποῦντες δ' εἰς ἀστραγάλων ῥυθμοὺς κομίζουσιν εἴς τινα νῆσον προκειμένην μὲν τῆς Πρεττανικῆς, ὀνομαζομένην δ' Ἴκτιν· κατὰ γὰρ τὰς ἀμπώτεις ἀναξηραινομένου τοῦ μεταξὺ τόπου ταῖς ἁμάξαις εἰς ταύτην κομίζουσι δαψιλῆ τὸν καττίτερον. [3] ἴδιον δέ τι συμβαίνει περὶ τὰς πλησίον νήσους τὰς μεταξὺ κειμένας τῆς τε Εὐρώπης καὶ τῆς Πρεττανικῆς· κατὰ μὲν γὰρ τὰς πλημυρίδας τοῦ μεταξὺ πόρου πληρουμένου νῆσοι φαίνονται, κατὰ δὲ τὰς ἀμπώτεις ἀπορρεούσης τῆς θαλάττης καὶ πολὺν τόπον ἀναξηραινούσης θεωροῦνται χερρόνησοι. [4] ἐντεῦθεν δ' οἱ ἔμποροι παρὰ τῶν ἐγχωρίων ὠνοῦνται καὶ διακομίζουσιν εἰς τὴν Γαλατίαν· τὸ δὲ τελευταῖον πεζῇ διὰ τῆς Γαλατίας πορευθέντες ἡμέρας ὡς τριάκοντα κατάγουσιν ἐπὶ τῶν ἵππων τὰ φορτία πρὸς τὴν ἐκβολὴν τοῦ Ῥοδανοῦ ποταμοῦ.

[1] *But we shall write about its [Britain's] customs and other special features in the part where we come to the campaign which Caesar undertook against Prettania. Here we shall describe how tin comes from there. Those who live around the promontory on the British island called Belerion are particularly hospitable, and because of their trading with other peoples have a civilised way of life. They prepare the tin by working the land which has it in an ingenious manner. [2] This land is like rock with strata of earth. In these they work a passageway, and clean it by melting. They cast it into pieces the shape of knuckle-bones and transport it to an island which lies off Prettania called Ictis; for at low tide the area in between becomes dry, and they take large quantities of tin to it in wagons. [3] Something special happens concerning the islands nearby which lie between Europe and Prettania. At full tide the passage between them and the mainland is full and they appear as islands; but at low tide the sea flows back and leaves a large area dry, and they look like peninsulas. [4] There [Ictis] the merchants buy [the tin] from the locals and transport it across to Galatia [Gaul]; finally they travel across Galatia on foot for some thirty days, bringing the goods on horseback to the mouth of the river Rhone.*

This text follows that printed at the end of the commentary on F4. The autopsy for both the tin area and the tidal islands must reflect Pytheas via Timaios for the reasons noted on F4; and, as proposed at p 7, we may reasonably infer that one aspect of Pytheas' voyage was to investigate the trade in both tin and amber.

§1 If Diodoros wrote about Caesar's invasions, it has not survived. As with the headlands in F4, the name of the island itself is from Pytheas. It is recorded as *Prettania* in the MSS of Diodoros (e.g. when naming it as the largest island in the Ocean just before F4) and by Ptolemy and the geographer Marcianus (F9), and in parts of Strabo; a scribe's correction to the more familiar *Bretannia* clouds the issue: Latin always spelled (and presumably pronounced) it with a B-, Britannia. The mainland population were P-Celts, and *Prettania* is regarded as correct: that is how Pytheas heard it and recorded it, reflecting the local pronunciation with a *p*- or perhaps *bh*-sound at the beginning.[1] As noted on p 126, it is clear that Pytheas did not know the name 'Albion'; or, if he did, he did not associate it with the island which he called *Prettania*; cf on F18. The Belerion promontory is Land's End (cf F4), and Cornwall would be Pytheas' first landfall after crossing from France.

As Pytheas sailed up the west coast of France, he would find the local Celts' life-style similar to what he was familiar with in Marseille; when he crossed to Cornwall, he found little difference, which he attributed to their trade contacts across the Channel. Hospitality to strangers was a basic obligation of Greeks; it was a compliment to note how 'Greek' these locals were. The word is ἐξημερωμένοι, *exēmerōmenoi*, 'particularly civilised', and it must come from Pytheas' description of being there. Diodoros perhaps included it in his epitome because it contrasted with the less civilised British lifestyle he had just described in 5.21.5, pp 32–3.

§2 The account of extracting tin is clearly autopsy, and must be from Pytheas. Behind the 'hospitable locals' of §1 and the account of its sale here and in §4, we may infer that they were happy to show him how they extracted the tin, but also made clear that they had well-established trade links across to the mainland and were not open to dealing directly with him or his Massiliot principals. It is not clear whether he appreciated that those who bought the tin were probably kin to his hosts, the sellers. The description of 'Ictis' makes it clear that it must be a local island just offshore, and not the Isle of Wight, for which the Roman name was Vectis (cf on F19). Whether the tin reached the coast on pack animals or carts, the use of carts to take the tin across to an island implies a sandy bay from which it had proved inconvenient to load ships. Taking the tin across to the island would be easier in an era of lower sea levels. The obvious candidate is St Michael's Mount, off the sandy bay east of Penzance.[2]

---

1 The occurrences are collected in Rivet and Smith (1979) 39–40 and discussed *ibid* 280–2, where the Q-Celt (Irish) 'Cruithin', their name for the Picts, is noted. His reference to the 'Massiliot Periplus' may be ignored: Appendix 1.

2 Cunliffe (2001) 78–9 urges the merits of Mount Batten in Plymouth Sound; but that is a longish way from Belerion/Land's End. In theory there is Burgh Island, Devon, since tin ingots, some arguably in

§3 This description of tidal islands interrupts the description of the tin trade. It perhaps refers to the Scillies. With lower sea levels, they would be one large irregular shaped island and some outliers. Their Roman name was always singular in its variant spellings: see on 'Silimnus' in F19. Although the text says that Pytheas was told about them, he might have passed and so seen them as he crossed from France to Cornwall, and asked about them on arrival. An alternative possibility is the Channel Islands, including Chaussey. They are now surrounded by shallow water and submerged rocks, but with lower sea levels the description could fit.[3] It is not Burgh Island (n 2): the text speaks of a group of islands. If Pytheas noted their name, it does not survive.

§4 There is a question as to the source of this paragraph. Cornish tin miners would know that their tin went across to Gaul, but how much they knew about its subsequent journey is uncertain. Pytheas perhaps recorded what he was told at Corbilo (F36), or learnt in Marseille from those bringing the tin there on the last leg of its journey.[4] A 30 day journey by pack animals from Corbilo is plausible; it would reflect daily stretches of 20+ miles, 32+ km.[5] If the tin was taken just across the English Channel, around Morlaix or at Lannion, the overland journey would be longer and less easily fit in to 30 days; but 30 is a round number, and some exaggeration by Pytheas' informants is conceivable. It is a matter of comment that he speaks of the mouth of the Rhone; Marseille was some 35 miles, 56 km, away by sea. It need not have been the only route: some merchants would prefer to take the tin by sea to Corbilo (F36) or to the Garonne at Bordeaux and overland from there (Bowen (1972) 58–60), probably a cheaper method. Also some of the tin reaching Marseille would have come from Brittany, and what was said about a tin route across France would apply also to that. Note that later in book 5, 5.38.5, from source(s) other than Timaios, Diodoros speaks of tin from Britain and Brittany being taken across Gaul on horseback to both Marseille and Narbonne.[6]

the knuckle-bone shape described here, were found in the estuary of the River Erme, a few miles to the west (Historic England List entry 1000054). But (a) that is even further away from Cornwall, and (b) the timber associated with the find is dated c 4000, and thought to be from a submarine forest and not a boat. Whatever the explanation for this tin, it would appear to be a red herring in the present context. Even allowing for lower sea levels, no other island on the Channel coast of Cornwall fits the bill; they are opposite cliffs and not accessible from a beach. The Scillies are too far away.

3 Google satellite view shows the present situation.

4 As noted on p 10, he probably had some idea of the distance from Marseille to the Ocean coast of Keltikē.

5 We can get an idea of the distance from the modern motorway route from Nantes to Marseille, 986 km, 616 miles; the motorways must reflect the sort of route pack animals would have taken.

6 Hawkes (1977) 31 suggests that Posidonios was Diodoros' source for the carriage of tin in F5, presumably on the basis that the former could have noted it when dealing with the Gulf of Lyon area. He could possibly have been the source for the later passage just noted; and while he is not usually thought to have been one of Diodoros' sources for book 5, the mention of Narbonne indicates a source other than Timaios. Cunliffe (2018) 161–2 includes a useful map suggesting the routes, partly maritime, partly overland, that would be used from time to time.

F6 Diodoros Siculus 5.23.1, 4 ≈ Timaios *FGrH* 566 F164.23

[1] Περὶ μὲν οὖν τοῦ καττιτέρου τοῖς ῥηθεῖσιν ἀρκεσθησόμεθα, περὶ δὲ τοῦ καλουμένου ἠλέκτρου νῦν διέξιμεν. τῆς Σκυθίας τῆς ὑπὲρ τὴν Γαλατίαν κατ' ἀντικρὺ νῆσός ἐστι πελαγία κατὰ τὸν ὠκεανὸν ἡ προσαγορευομένη Βασίλεια. εἰς ταύτην ὁ κλύδων ἐκβάλλει δαψιλὲς τὸ καλούμενον ἤλεκτρον, οὐδαμοῦ δὲ τῆς οἰκουμένης φαινόμενον. περὶ δὲ τούτου πολλοὶ τῶν παλαιῶν ἀνέγραψαν μύθους παντελῶς ἀπιστουμένους καὶ διὰ τῶν ἀποτελεσμάτων ἐλεγχομένους . . . [4] . . . τὸ γὰρ ἤλεκτρον συνάγεται μὲν ἐν τῇ προειρημένῃ νήσῳ, κομίζεται δ' ὑπὸ τῶν ἐγχωρίων πρὸς τὴν ἀντιπέρας ἤπειρον, δι' ἧς φέρεται πρὸς τοὺς καθ' ἡμᾶς τόπους, καθότι προείρηται.

[1] *But as regards tin we will be satisfied with the above, and we shall now deal with what is called electron. Opposite the part of Scythia which is beyond Galatia [Gaul], there is an island in the Ocean called Basilia. On it the waves throw up what is called electron in great quantities, appearing nowhere else in the inhabited world. Many old writers have written up totally unbelievable stories about it, disproved by later events. . . . [4] For amber is collected on the island mentioned above, and it is taken by the locals to the mainland opposite, from which it is carried to our area, as has been said.*

This text follows F5, and the factual parts are still based on Timaios using Pytheas. 'Electron' meant both an alloy of gold and silver, and amber. By adding 'what is called', Timaios (or Diodoros) made clear that it meant amber. European amber reached the Mediterranean from two areas: the west coast of Jutland, and the Gulf of Danzig (Gdansk) and the coasts of the Baltic states, Samland. It is the fossilised sap of pine trees, and on these coasts the marine sands are rich in strata containing it, being gradually released by the action of waves. This was unknown in the ancient world, and there were numerous imaginative explanations for it. Pliny collected them when discussing amber: see on F21. Traders spoke of it coming from distant islands; and it is interesting that when in the Baltic Pytheas found it only on an island, and never discovered its real origins, the mainland. A small quantity is washed up on islands (including stretches of the English east coast: cf F21).[1]

Both Pytheas' and Timaios' conception of northern Europe was that Keltikē merged eventually into Scythia, but with no idea where, or that the Rhine existed, or that German tribes lived somewhere east of Keltikē: see pp 9–10 with map 1, and Appendix 4. So for Pytheas, his amber island was off the coast of Scythia. 'Basilia' cannot be the local name of an island inhabited by non-Greek speakers. It is a pure Greek word: 'queen' or 'kingdom' as a noun, 'royal' as an adjective. We have other references to the island and its name in Pliny, F17 and F21. They are considered

---

1 Nowadays most amber is mined, but in Greek and Roman times it is probable that significant quantities were washed up; Pliny reports that around 60 AD a Roman *eques* had been able to procure a huge supply of Baltic amber with no difficulty (*NH* 37.45).

in Appendix 4, with the conclusion that the local name was *Balchia* ('ch' as in 'loch') or *Baltsia*, or similar, which Pytheas recorded as the euphonious 'Basilia'; and that it was located in the Baltic, probably the Denmark area. Pytheas' version of the name is no evidence that the island was ruled by a king or the amber trade was a royal monopoly, as has been proposed, e.g. by Hawkes (1977) 9–10).[2]

Diodoros was writing at a time when Caesar was conquering, or had conquered, Gaul, but probably before the latter had written his *Bellum Gallicum*. He would know of the information that was filtering back to Rome from Gaul: he can say that Keltikē is 'now known as Galatia' (Gaul in Latin, 4.19.4, in the context of the battle of Alesia, 52), and can add that it extends to the Ocean (5.21.1, just before F4). He knew that the Rhine existed (5.24.4), but says that Caesar was attacking Celts in the far bank; the 'German' name does not appear, at least in the surviving parts of his book. At all events, it is doubtful whether he could have offered a better location for the amber island than 'Scythia'.

The phrase 'as has been said' in §4 must relate to the transport of tin at end of F5: amber too was carried overland by pack animals. Pliny *NH* 37.43–5, within his discussion of amber, deals in some detail with an overland route for amber from the Baltic to Carnuntum by Germans,[3] and thence by the Veneti to the Adriatic. Roman Carnuntum only dated from about 6 AD, but it must reflect a long-established trading route, with the Danube a major exchange hub. Amber from Jutland will mostly have reached Marseille through France.

In the light of Herodotos' mention of an Eridanos river and amber (p 8), we need a note about the 'unbelievable stories'. The passage omitted relates the myth of Phaëthon, son of Helios, the sun god, in this version killed when he drove his father's chariot too near the sun. He fell to the ground at the mouth of the river Po, 'formerly called the Eridanos'; his sisters were changed into poplar trees, and their tears hardened into amber.[4] Pliny includes a similar version in his collection of amber stories (*NH* 37.31, a page or so before F21), naming the poets who tell it. Diodoros adds that the story is false, because amber is not found in the Adriatic; but the Po was a very old trading route, as evidenced by the late Bronze Age trading centre of Frattesina on a former branch of the Po, some 40 to 50 miles inland (64 to 80 km), and, as just noted, amber for the Mediterranean went to the Adriatic. See also on Glaesiae in F19.

---

2 See e.g. the discussion in Roseman (1994) 102. No doubt there was a local chieftain; Pytheas would meet them everywhere. Even if the chieftain on this island was also the man who ran the local amber trade, he was not saying that; he was only recording a euphonious version of a local name in another language.
3 Its extensive remains are at Petronell-Carnuntum and Bad Deutsch-Altenburg on the Danube in Lower Austria, only a few miles from the frontier with Slovakia, about 15 miles from Bratislava.
4 Diodoros' Greek has a subtlety lost in translation: the trees give off δάκρυον, *dakryon*, which in the singular, as here, means either 'tear' or 'sap'. That detail would not be inaccurate if it related to a different species of tree. For the variant versions of the myth, such as whether Phaëthon was driving the chariot with or without his father's permission, and the different places where it happened, see Gantz (1996) 31–4.

F7 Geminos 6.7–9

[7] Οὐ κατὰ πᾶσαν δὲ χώραν καὶ πόλιν τὰ αὐτὰ μεγέθη τῶν ἡμερῶν ἐστιν. Ἀλλὰ τοῖς μὲν πρὸς ἄρκτον οἰκοῦσι μείζονες αἱ ἡμέραι γίνονται, τοῖς δὲ πρὸς μεσημβρίαν ἐλάττονες. [8] Ἔστι δὲ ἐν Ῥόδῳ μὲν ἡ μεγίστη ἡμέρα ὡρῶν ἰσημερινῶν ιδ∠, περὶ δὲ Ῥώμην ἡ μεγίστη ἡμέρα ὡρῶν ἰσημερινῶν ιε· τοῖς δὲ βορειοτέροις οἰκοῦσι τῆς Προποντίδος ἡ μεγίστη ἡμέρα γίνεται ὡρῶν ἰσημερινῶν ις, καὶ ἔτι τοῖς βορειοτέροις ιζ καὶ ιη ὡρῶν ἡ μεγίστη ἡμέρα γίνεται. [9] Ἐπὶ δὲ τοὺς τόπους τούτους δοκεῖ καὶ Πυθέας ὁ Μασσαλιώτης παρεῖναι. Φησὶ γοῦν ἐν τοῖς Περὶ τοῦ Ὠκεανοῦ πεπραγματευμένοις αὐτῷ ὅτι ἐδείκνυον ἡμῖν οἱ βάρβαροι ὅπου ὁ ἥλιος κοιμᾶται· συνέβαινε γὰρ περὶ τούτους τοὺς τόπους τὴν μὲν νύκτα παντελῶς μικρὰν γίνεσθαι ὡρῶν οἷς μὲν β, οἷς δὲ γ, ὥστε μετὰ τὴν δύσιν μικροῦ διαλείμματος γινομένου ἐπανατέλλειν εὐθέως τὸν ἥλιον.

[7] *The length of days is not the same in every land and city. For those living in the north the days are longer, in the south shorter.* [8] *In Rhodes the longest day is 14½ equinoctial hours; in Rome, 15. For those living to the north of the Propontis the longest day is 16 equinoctial hours; further north still it is 17 and 18 hours.* [9] *It is to these regions that Pytheas of Marseille seems to have come. At any rate he says in his book* On the Ocean *that the locals showed us where the sun sleeps; for it happens in these regions that the night becomes very short, in some places 2, in some 3 hours; so that after sunset there is a short interval and it immediately rises again [trans Evans and Berggren (2006), adapted].*

Geminos was a mathematician and astronomer of the first century who wrote an elementary textbook on astronomy, its exact title being uncertain.

§§7–8 These daylight figures broadly correspond to those offered by Hipparchos, for which see on F28 and Appendix 2 §§5–7; cf Cleomedes' figures noted on F2. Hipparchos' 15 hours was located 'between Rome and Naples'; Geminos' 'Rome' may be compression, or from another astronomer's later revision; it is accurate for either city.[1] 'Equinoctial hours' meant those measured on a sundial calibrated at an equinox: the hours were of equal length throughout the year and at any place: see p 137. Calibration could only be done in daylight; the Greek literally means 'equidiurnal' (equinox is ἰσημερία, *isēmeria*, 'equal days').

§9 There are three important details here. Firstly, the 'us' in 'showed us' is good evidence that Pytheas was sailing with a crew (cf p 7). Secondly, it offers the title of his work, as in F3; see on F22 and F30 §6 for the alternative title in F22. The third point turns on how to punctuate the rest of the sentence. Did Pytheas say 'the locals . . . sleeps', the rest being Geminos' comment; or is the whole 'the

1 Modern figures are 15 hrs 14 min for Rome and 15 hrs 7 min for Naples.

locals . . . again' from him? Editors of both Pytheas and Geminos are divided.[2] In favour of the whole text, Bianchetti (1998) 191 notes (a) that the word for 'it happens' (συνέβαινε, *synebaine*) is imperfect, which 'underlines the continuity' with what precedes, what the locals were telling him, and explains what follows; and (b) Geminos is reporting lengths of days; a very long day and very short night would be a novelty for Greeks, which Pytheas would want to explain. For the shorter sentence, Roseman (1994) 143–3 stresses that there was no method or equipment for measuring the length of nights in Pytheas' time (as we have also seen on F2), and even Hipparchos' table of day lengths was from observing the sun's shadows, i.e. during daylight, and mathematical calculations; so the 2 or 3 hours was an explanatory comment by Geminos.

None of these views touch on Geminos' use of ὥρα, *hōra*, in the modern sense of one twenty-fourth of a day. In Pytheas' time, it did not have that meaning. It meant the time for doing something, or any period of time: a day, a season, a year. To subdivide a day, Greeks used expressions such as 'early' or 'when the agora is filling up';[3] in the evening they could sometimes refer to the length of your shadow: Aristophanes makes a wife tell her husband not to come home until his shadow is ten feet long.[4] It was only after sundials were developed that *hora* was adapted to refer to the divisions of daylight, and so its modern meaning. This is discussed in Appendix 3, with the conclusion that sundials were a novelty in Pytheas' time, and even if *hora* was already the astronomers' word for the divisions marked on them, it had not yet passed into general circulation and would not have been understood by his readership.[5] By Geminos' time, the word did have its modern meaning, and the 2 and 3 hours here is almost certainly Geminos either 'translating' Pytheas into more modern language or his own comment.[6]

2 E.g. Pytheas: Mette (1952) 28 just the first sentence; Bianchetti (1998) 103 the whole; Roseman (1994) 139–40 prints the whole, but her commentary, 143, considers Mette correct. Geminos: Manitius' Teubner, the first sentence; Aujac's (1975) 34–5 the whole; Evans and Berggren (2006) 162 the whole, with a footnote referring to Roseman.
3 Πρωί, *proï*, early; μεσημβρία, *mesēmbria*, midday; ὀψέ *opse*, late; πλήθουσα ἀγορά, *plēthousa agora*, the agora filling; ἀγορὰ λυθῆναι, *agora luthēnai*, the agora breaking up (literally 'loosing'). As instances, Hdt 8.6.1, περὶ δείλην πρωίην, *peri deilēn proïēn* 'about early afternoon'; 9.101.2, πρωί . . . περὶ δείλην, *proï . . . peri deilēn*, 'early in day . . . in the afternoon'.
4 Ar *Eccl* 652; other shadow times Ar F695, Men F625, Euboulos F117. While these are all from comedy, the shadow time in IG xii.5 647.16 shows that it was not a joke (even if the 20 feet in Eubulos is). All references are to dinner time, and perhaps this usage was limited to the evenings. Generally Sommerstein (1998) 196–7, fuller than Ussher (1973) 166–7. The water-clock of the law court was based on so many liquid measures (*chous*), or, for public suits, the length of day in Poseidon (roughly December), for which see Arist *Ath Pol* 67.3–4 with Rhodes (1993) *ad loc*.
5 Neither *LSJ* nor *CGL* make clear that *hora* as 'hour' dates from and relates to the development of the sundial.
6 The 'sun's bed' was also noted in F3. Latin *hora* does not help; by the time of our earliest texts, late third century, it could mean an hour: Plautus fragment, *Boeotia* F1.2; Terence Eun 341, Phorm 514.

However we read this passage, when and where did Pytheas learn it? It would be in the Northern Isles, and in the winter. If he was there in the summer, he would experience short nights and see for himself where the sun set and then rose. On the other hand, it is reasonable to infer that he overwintered in the Northern Isles, probably the Shetlands. He would reach the Northern Isles at the end of his first year's sailing season: see the Coda p 174. He then experienced long nights and short days, and the locals will have pointed to the north-north-west for where the sun set in summer and to the north-north-east for where it soon rose. In the Shetlands there are some 6 hours of daylight even at the winter solstice.

F8 Hipparchos of Nicaea, *Commentary on the* Phaenomena *of Aratos and Eudoxos* 1.4.1

περὶ μὲν οὖν τοῦ βορείου πόλου Εὔδοξος ἀγνοεῖ λέγων οὕτως· ἔστι δέ τις ἀστὴρ μένων ἀεὶ κατὰ τὸν αὐτὸν τόπον· οὗτος δ᾽ ὁ ἀστὴρ πόλος ἐστι τοῦ κόσμου. ἐπὶ γὰρ τοῦ πόλου οὐδ᾽ εἷς ἀστὴρ κεῖται, ἀλλὰ κενός ἐστι τόπος, ᾧ παράκεινται τρεῖς ἀστέρες, μεθ᾽ ὧν τὸ σημεῖον τὸ κατὰ τὸν πόλον τετράγωνον ἔγγιστα σχῆμα περιέχει, καθάπερ καὶ Πυθέας φησὶν ὁ Μασσαλιώτης.

Concerning the north pole Eudoxos is wrong when he says this: there is a certain star which always remains in the same place, and this star is the pole of the world. For there is no one star at the pole; it is an empty place, around which three stars turn. The boundary of these encloses a shape at the pole almost like a tetragon, as Pytheas of Marseille also says.

This comes from the only surviving work of the astronomer Hipparchos (p xx). Eudoxos, as noted on p 19, wrote a description of the heavenly bodies called *Phaenomena*, 'things seen', and Aratos, a writer with philosophical training of the early second century, wrote a poem based on it. Hipparchos then wrote a critical commentary on both works.[1] The statement about stars around the north pole is correct. They are part of the constellation of Ursa Minor; both then and now there was and is no actual star at the north pole, but a blank area. But there is always one bright star near to the pole which is treated as the pole star. Nowadays it is α Ursae Minoris, 1° away; in Eudoxos' and Pytheas' time that star was more than 12° distant;[2] their pole star was β Ursae Minoris.[3]

---

1 Both Aratos and Hipparchos are important sources for recovering Eudoxos' works, though there is a question how far Aratos' poem more or less put Eudoxos' text into verse and how far it differed. That does not affect this reference, which solely concerns Eudoxos. The scope of Eudoxos' work is noted at pp 19–20.
2 Hipparchos recorded it some 170 years later as 12° 24′ (Ptol Geog 1.7.4).
3 See the maps in Roseman (1994) 147 and at the end of Bianchetti (1998); explained, Dicks (1960) 170.

Eudoxos' error is curious. Lasserre (1966a) 187 notes Anaxagoras saying that the pole star used to be directly above you, and then tilted; but that was just speculation.[4] Even in a world where astronomy depended solely on what could be observed with the naked eye, Eudoxos would know what he could see above his head; and if Aratos vv 24–30 reflects him, he knew perfectly well that the two bear constellations moved round an empty space. Indeed, you did not need to be an astronomer to know your constellations. Any peasant farmer had a working knowledge of the stars, for instance when constellations rose and set, as an aid to knowing when to sow or harvest;[5] mariners would also know their constellations, and steering by the pole star was an ancient technique.[6] A possible explanation for his 'error' may lie in another of Eudoxos' works: *On Speeds*, which for the first time described mathematically the apparent motions of the heavenly bodies, including the planets, as a series of circles and concentric circles, from the viewpoint of an observer on the earth, stationary in the middle of the celestial globe. It was a very considerable achievement, and although later writers could correct it, it was, for its time, a major advance. He may there have used β Ursae Minoris as a suitable fixed point for one end of the diameter of one of his circles, or through which one of his circles ran. Against that, and even without the Aratos citation, (a) we would expect his *Phaenomena* to be visually accurate, and (b) Hipparchos would have known *On Speeds*, and therefore whether the pole star was being used as a fixed point.

That problem apart, there are two interlinked questions concerning Pytheas: is the comment that Eudoxos was wrong his or Hipparchos', and in what context did he mention the pole star? Pytheas had a good education in astronomy, pp 19, 22, and was also a mariner. The text reads more easily on the basis that it was Hipparchos who said that Eudoxos was wrong, citing Pytheas only for the tetragon shape. As to the context, it is unlikely to be part of a longer passage stating which constellations were or were not visible as he travelled north. As noted in Appendix 2 §5, Hipparchos' definitions of some latitudes included which stars were visible; but it is doubtful if the astronomy of Pytheas' time understood the correlation between visible constellations and latitudes: see on F2 with n 9. It is feasible that he had a difficult overnight voyage, hampered by fog or mist, and said that he could only

---

4 Lasserre refers to Anaxagoras 59 A42 D–K = Hippol *Comm in Arat* 1.8.6, that all the heavenly bodies are included in the rotation of the aether; cf 59 A1 D–K = Diog Laert 2.9, the pole star being originally vertical and then tilting; Guthrie (1965) 305–6; Dicks (1970) 59–60. The matter is complicated by the fact that the word 'pole' had a number of meanings, including both the celestial dome and an axle, including a celestial axle.

5 We have useful insight into this from the many references to constellations in Hesiod's agricultural *Works and Days*: Pleiades, 383–8; Sirius, 417–19, 587; Arcturus 571–3; Orion 597–9; Orion, Sirius, Pleiades, Hyades, Orion, 609–17; Pleiades 618–21.

6 Night sailing was common, and the stars at the pole were of particular interest to mariners. As an example, in the fourth century *Periplous* of Ps-Scylax, there are 31 distances between two ports in terms of day and night sailing, reflecting long standing knowledge, e.g. sec 7: Sardinia to Libya, a day and a night; Sardinia to Sicily, two days and nights.

rectify his course once the fog cleared and he could see the pole star. If he noted the pole star as part of a longer passage on how different constellations, or parts of them, did or did not become visible as he moved north, only this passage survives; unless we can also attribute to him the reference to Cancer, the Crab, in F2. But, as there noted, it is doubtful whether we can extend Cleomedes' statement that others put Pytheas at Thule to saying that the Cancer reference was also his. No other reference even hints that Pytheas mentioned constellations.

F9 Marcianus of Heracleia, *Epitome of the Periplus of the Inner Sea by Menippos* 1.2, *GGM* 1.565

οἱ γὰρ δὴ δοκοῦντες ταῦτα μετὰ λόγων ἐξητακέναι, Τιμοσθένης ὁ Ῥόδιός ἐστιν, ἀρχικυβερνήτης τοῦ δευτέρ[αι]ου Πτολεμαίου γεγονώς, καὶ μετ' ἐκεῖνον Ἐρατοσθένης ὃν Βῆτα ἐκάλεσαν οἱ τοῦ Μουσ<ε>ίου προστάντες, πρὸς δὲ τούτοις Πυθέας τε ὁ Μασσαλιώτης καὶ Ἰσίδωρος ὁ Χαρακηνός . . .

Those who have seen fit to examine these things (i.e. geography) with reasoning include Timosthenes of Rhodes, naval commander under Ptolemy II, and after him Eratosthenes, called beta by those in charge of the Museum; also Pytheas of Marseille and Isidoros of Charax.

Marcianus cannot be dated more accurately than between 200 and 530 AD; he published his own *Periplus of the Outer Sea* (in fact from the Arabian Gulf to the Indian Ocean), and epitomes of two earlier works of the first century BC: one of Artemidoros' (p xix) *Geography*, and one of a *Periplus* by Menippos, from which this text comes. Menippos himself probably dates from the mid first century; Strabo cites him twice.[1] As noted on p 7 n 1, *periplous (periplus* in Latin), 'sailing round', and *periodos gēs*, 'journey round the world', 'world guide', were the titles of a large number of travel guides or sailing manuals; they go back to the earliest days of written texts, from Scylax of Caryanda in the late 6th century (n 7) onwards. They typically went from harbour to harbour, with practical details such the distance between them, if there was good anchorage, and noting an important landmark such as a temple on a nearby headland; that was often accompanied with the local myth associated with its dedicatee: the mariner could give thanks for his safe arrival to the god or hero of the temple.[2] It is against that background that we should assess this reference to Pytheas.

Our text is from the introduction, where Marcianus says that he has read many *periploi*, and selects 18 for their quality. The Greek, μετὰ λόγων, *meta logōn*, is

---

1 The title of his work is unknown, but is usually cited as here, *Periplus of the Inner Sea.*
2 Headlands were an essential tool in ancient navigation; cf the reference to them at p xviii. Churchill Semple (1927) has a very useful review of headlands with temples in the Mediterranean and Black Sea; see the maps at 354 and 356.

difficult to translate but has the sense of reasoning, relying on solid facts.[3] The list includes books with other titles (Pytheas' was *On the Ocean* (F3, F7)); some we could call geographies. They mostly cover the Mediterranean and Black Sea (the 'Inner Sea' of the title), or the Red Sea and Indian Ocean. Two dealt with the Ocean: Pytheas for the area north of Gibraltar, and Hanno of Carthage (p 13), for the west coast of Africa.[4]

The list is not chronological. Timosthenes lived in the mid third century; his office, here translated 'naval commander', is literally 'chief steersman' or 'chief navigator'. His work survived for a long time because of its practical value;[5] it increased Aristotle's wind rose (*Meteor* 363a 21-365a 13) from 10 winds to 12, and became a standard for mariners. Timosthenes' contemporary Eratosthenes wrote a geography, not a *periplus*; for his 'beta' epithet, see p xx. For Isidoros of Charax (end of first century), see on F18. Of the other 14, only one survives in full, Strabo;[6] Hanno survives in a possibly condensed Greek translation (p 13 n 24). For some we only have fragments, including Euthymenes, p*[c3], and some are little more than names to us.[7] If, which is at best uncertain, Philemon wrote about the north from autopsy (p xx), his book did not merit inclusion.

Pytheas' inclusion is important for three reasons. It confirms that the hostility of Strabo and Polybios was not shared by others; secondly, that copies of his work still circulated as late as Marcianus' time. Thirdly, the expression 'with reasoning', true where we can check the others whom he names, shows how much we have lost of Pytheas, as discussed in the Coda, pp 180–3.

F10 Martianus Capella, *De Nuptiis Philologiae et Mercurii* 6.595 Préaux

Hinc est quod in Meroe longissimus dies duodecim aequinoctiales horas et alterius bessem secat, Alexandriae quattuordecim, in Italia quindecim, in Britannia decem et septem. solstitiali vero tempore cum caeli verticem sol invectus subiectas deorsum terras perpetui diei continuatione collustrat. itemque brumali descensu,

---

3 The phrase occurs in a comparable sense at Plato *Rep* 548b8–c1. Bianchetti (1998) 107 has [those] 'che sembrano essersi dedicati con raziocinio', 'who seem to have dedicated themselves with reasoning'.
4 See on F22 for rejecting the suggestion that Pytheas wrote about the Inner Sea, the Mediterranean.
5 For instance, he noted a temple at the Black Sea entrance to the Bosporos (Harpocrat *Lex* 143, Schol Ap Rh 2.531–2), and an island Artake just off the coast at Cyzicos offering a good anchorage (Herodian Pros Cath 3.1.314). Pliny often quoted his distances.
6 One wonders what Strabo would have said if he had known that after his death, his name would be coupled with Pytheas' on an equal footing.
7 Menippos' criteria for inclusion can also be judged from his noting that Scylax and an otherwise unknown Botthaios (possibly a misspelling for Hecataios) specified distances in days' sailing and not stades, practical details for a navigator. The surviving *periplous* attributed to Scylax is a later compilation, but contains some material reflecting the original.

semiannuam facit horrere noctem, quod in insula Tyle compertum Pytheas Massiliensis asseruit. his temporum diuersitatibus assertum, ni fallor, globosam rotunditatis flexibus habendam esse tellurem.

As a result at Meroë the longest day lasts for 12 equinoctial hours and ⅔ of another, 14 hours at Alexandria, 15 in Italy, and 17 in Britain. Indeed, at the time of the [summer] solstice, when the sun has reached the top of the sky, it lights up the lands below with continual perpetual daylight. Similarly at its winter descent, it makes six months of shivering night, which Pytheas of Massilia asserted he found in Thule island. For these differences in times it has been asserted, unless I am mistaken, that the earth is to be considered spherical, on account of the variations in its circumference.

Martianus, fifth century AD, was a native of Carthage. This text and F11 come from a curious encyclopaedic work in 9 volumes, partly in verse; its title ('The Marriage of Philology and Mercury') was given a little later by Fulgentius, a writer also from North Africa. It postulates that Mercury seeks a wife. He rejects Wisdom, Divination and Soul, and is advised to wed Philology. She is then made divine and given seven maidens to serve her, each one personifying the seven liberal arts (books 1–2). Books 3 to 9 describe each art in turn; book 6 deals with geometry (literally 'measuring land' and including geography). The work has no merits for originality; this text has been rewritten from Pliny: see on F15, where the substance is discussed.[1] The slight variations in wording do not encourage us to think that Martianus also used another reference to Pytheas. It is unclear how far he understood the science. He used Geminos for F11, but here overlooked a text such as Gem 6.13–15 noted in Appendix 2 n 7. For equinoctial hours, see on F7.

The Latin of the final sentence is obscure, but the meaning is clear: the lengths of daylight increase with the latitude of the place, the one being a function of the other. Stahl *et al* (1977) 223 translate, 'those discrepancies of the seasons . . . compel us to admit that the earth is round'. Gasparotto (1983) offers 'date le incurvature della sua rotundità', 'given the curvatures of its roundness'; Bianchetti (1998) 99 'a causa delle varazioni di circonferenza' ('circumferences' meaning latitudes), which we have adopted here.[2] Bianchetti 179 points to Arist *De Caelo* 297a–298a: one of his arguments for the earth being a sphere is that you see different stars at different latitudes; but Martianus is here citing Pytheas for different lengths of daylight.

---

1 Roseman (1994) 80 says that the earth being a sphere had relevance for Martianus (who is thought to have been pagan) because there was a question whether that was incompatible with scripture. As noted on F3, that was always a minority Christian view.
2 Roseman, n 1, cannot be right with her '[round earth] with spherical rotations'; that the earth spins on its axis was unknown.

F11 Martianus Capella, *De Nuptiis Philologiae et Mercurii* 6.608–9 Préaux

[608] . . . stellae etiam fixae caeli sex <eis> videntur mensibus, sex itidem non apparent, ortivusque circulus aequinoctialis illis est, senaque ex zodiaco signa conspiciunt; denique sex mensibus diesque noctesque patiuntur, ut utrisque poli axisque termini supra verticem videantur. sed haec prior Septentrionis, altera Canopi stellae illustrata fulgore cetera non noverunt. [609] quarum regionum habitus prodidit doctissimus Pytheas, sed ego ipsa peragravi, nequa mihi ignota videretur portio superesse telluris.

[608] . . . *[for those living in the icy north and south] the fixed stars are visible for six months, and again are not visible for six; the equator is the circle that marks risings for them, and six signs of the zodiac are visible; then they endure six months of night, so that for each the poles and the ends of the [earth's] axis are seen. The one [region] is illuminated by the light of the Great Bear, the other by that of Canopus, and they do not know of other [constellations]. [609] The most learned Pytheas reported the situation of these regions, but I have travelled through there myself, so that no part of the earth might remain unknown to me.*[1]

For Martianus and his work, see on F10. This text comes a page or so later, and is copied from Pliny *NH* 2.177–9, and Geminos (for whom see on F7) 5.31–4. As with the end of F10, it is a hard passage to translate; the above owes much to Stahl *et al* (1977) 227 and Gasparotto (1983) 63. Martianus has referred to the five zones into which the earth is divided horizontally, as proposed by Aristotle (p 20), and now notes both the polar regions. He could not know that he was correct about daylight and night for someone living at the north or south pole itself. He is not here referring to an arctic circle, either in the ancient (p 28) or modern sense; he claims that only the Great Bear is visible at the north pole during its six months' night, and Canopus at the south pole.[2]

The reference to Pytheas depends on restoring his name in the text from corrupt MS readings, but it is generally now accepted.[3] His placing Pytheas near the north pole shows that either he did not fully understand the science (cf Stahl *et al, loc cit* n 48), or he misunderstood his sources.

His statement that Pytheas 'reported' ('prodidit') the position is not evidence that he had Pytheas' book, even if copies still existed in his time: he was at least two or three generations younger than Marcianus, F9. He was merely copying Pliny,

---

1 The Latin is not easy to translate, and largely depends on Stahl *et al* (1977) and Gasparotto (1983).
2 Canopus was known as a constellation that began to be visible towards the southern limit of the habitable zone.
3 See Roseman (1994) 111–12. One MS has 'Pytharas' and the others 'Pyt(h)agoras'. The restoration is due to Préaux' 1969 Teubner edition. Older editions printed 'Pythagoras', but the concepts of long days and nights in the far north were quite unknown to him, and indeed at all before Pytheas.

F15. The 'most learned' epithet is also no more than an inference; a man who could write about such things had to be clever. It could only be a refutation of others' hostility if we assumed that Martianus was a history enthusiast who had read widely, including Polybios, and recalled the one paragraph where the latter attacks Pytheas, F30, or that he was familiar with books 1 and 2 of Strabo's massive *Geography*.[4] The 'I' who visited the north is the maiden personifying Geometry (cf on F10), as is clear from the feminine 'ipsa'.

F12 Mela 3.47–50

[47] . . . in Celticis aliquot sunt, quas quia plumbo abundant uno omnes nomine Cassiteridas adpellant. [48] Sena in Britannico mari Ossismicis adversa litoribus, Gallici numinis oraculo insignis est, cuius antistites perpetua virginitate sanctae numero novem esse traduntur . . . [49] Britannia qualis sit qualesque progeneret, mox certiora et magis explorata dicentur. quippe tamdiu clausam aperit ecce principum maximus, nec indomitarum modo ante se verum ignotarum quoque gentium victor, propriarum rerum fidem ut bello affectavit, ita triumpho declaraturus portat. [50] ceterum ut adhuc habuimus, inter septentrionem occidentemque proiecta grandi angulo Rheni ostia prospicit, dein obliqua retro latera abstrahit, altero Galliam altero Germaniam spectans, tum rursus perpetuo margine directi litoris ab tergore abducta iterum se in diversos angulos cuneat triquetra et Siciliae maxime similis . . .

[47] *On the Celtic coast there are some [islands], which, because they produce a great deal of tin, are together called the Cassiterides.* [48] *Sena in the British sea, opposite the shores of the Ossismici, is noted for its oracle of a Gallic god, and its priestesses, said to be nine in number and sanctified by perpetual virginity.* . . . [49] *As to what Britain is like and the people it produces, this will soon become definite and better explored and will then be spoken of; because, though closed off for a long time, the greatest of emperors, victor of previously unsubdued and quite unknown tribes, is opening it up. Just as he has striven for proof of his achievements in war, so he brings them to declare in his triumph.* [50] *But, as we have held up to now, it projects in a large obtuse angle between the north and the west so as to look towards the mouths of the Rhine. Next it draws back its sides obliquely, one facing Gaul, the other Germany; finally, drawn back on its rear side in a continuous line of a straight shore, it forms a three-sided wedge shape with its opposite angles, very much like Sicily.*

This text and F13 come from Pomponius Mela's *Chorography*, the first geography in Latin. Books 1 and 2 cover the Mediterranean; our text comes from book 3, dealing with Europe and Asia. His dates can be fixed from §49 above: an emperor

4 The eventual publication and knowledge of Strabo's *Geography* is a vexed question: Diller (1975) 3–12.

is about to have a triumph for his British victory. Its extravagant terms leave it open whether he refers to Caligula, who is said to have planned to invade Britain, or Claudius, who did invade in 43 AD and had a triumph in 44 AD; the planning for it would be known beforehand. On any view, we may put Mela between 37 and 44 AD.[1] The paragraph is diplomatic (or sycophantic) nonsense; apart from Pytheas, Timaios and Diodoros had written about it (F4–F6), and Julius Caesar both invaded and wrote about it; Mela himself goes on to say quite a lot about Britain, as noted below. But that was largely merchants' tales, not autopsy; he was writing before any details about most of Britain could circulate from Roman military or civil sources. Having dealt with mainland Europe, from the Straits of Gibraltar to the Baltic, he turns to the islands: first, Cadiz, and then the mythical Erythia (cf on F34; Mela places it in Lusitania, Portugal). Our text comes next. The following commentary should be read with map 2, p 51.

§47 The Cassiterides, the tin islands, never existed in fact, but belief in them persisted even after it was known that tin came from mines on the mainland; as noted on pp 8–9 it is futile to identify them with actual islands.[2]

§48 Sena, in the 'British sea', not noted elsewhere, is usually identified with the Ile de Sein, opposite the Pointe du Raz in Brittany, given the phonetic similarity. The Ossismici are the Breton tribe discussed in Appendix 5. It seems that various stories about Celtic priestesses on an island circulated: Mela describes his as shape-shifters with magic powers to create storms and cure diseases. They could be compared to a story Posidonios picked up about an island at the outlet of the Loire on which only women lived who worshipped Dionysios with mystic rites; that is of some interest here because they were probably of the Namnetes tribe (cf F36).[3]

§49 This eulogy of the emperor has been noted above.

§50 Mela moves on to Britain. He starts with the east coast, which for him runs from south-west to north-east, and it 'looks towards' (*prospicit*) the Rhine; it is more vague, though from our perspective more accurate, than Strabo's

---

1 Caligula, emperor 37–41 AD, had spoken of invading Britain, though it is not clear what he actually did; our basic information is Cassius Dio 80.20–23. For the problems of dating Mela see Romer (1994) 2–3; Silberman 282, who accepts 43–4 AD. Romer is probably correct in seeing a pun on Claudius' name; Mela's 'closed off', is *clausa*, from the verb *claudo*, shut. He must have published around this time, because Pliny names him as one of his sources for many of his books, including his book 4, *NH* 1.4.

2 Spain and Gaul were under Roman rule in Mela's time, but it does not follow that he knew that some tin came from there. If he knew what Pytheas or Timaios or Diodoros had written about Cornish tin (F5), it did not overcome his innate belief in tin islands. In any case, Cornwall was unknown to the Romans in 43–4 AD (except to a shrewd reader of F5).

3 Posid F276 E–K = Strabo 4.4.6. They are called 'Samnites' in the MSS, but these were an Italian tribe, and it is reasonably certain that the story was about the Namnetes, as in F37, Caesar *BG* 3.9.10 and Pliny *NH* 4.107: see Kidd (1988) 939. There is no island in or near the outlet of the Loire. Belle-Ile-de-Mer and the two smaller islands nearby, Houat and Hoédic, are well over 30 miles away.

'opposite' the Rhine in F25. The south coast is opposite Gaul, and the west coast ('rear side') also runs from south-west to north-east, so that Britain is an isosceles triangle leaning to the right. The 'obliquely' reference goes back to Pytheas, but the isosceles triangle orientation probably does not: see on F4. A page lower down, at 3.53, he puts Ireland lying 'super', 'above', i.e. north, of Britain. His northern geography is further considered on F13. Silberman (1988) 282–3 proposes Eratosthenes as Mela's source. Eratosthenes used Pytheas for Britain, including its three sides and (probably) its comparison with Sicily: see on F4. But the reference to the Rhine must be from some later geographer; Pytheas did not record it, so far as we know, and it was unknown until much later (Note to Appendix 4, introduction and §1). Strabo's view of Britain was different, as discussed in Appendix 7. Mela's description of Britain that follows this text (down to 3.52) cannot depend on Pytheas to any extent; it says that the island is flat (*plana*); if Pytheas described it physically, he must have mentioned mountains. Mela says that the island is fertile and good for sheep; its inhabitants dye their bodies blue and are uncivilised; in war they fight with chariots with scythed wheels; and they have rivers, some flowing into the sea, some back on themselves, 'some of which produce gems and pearls'.[4] It may be compared to Diodoros 5.21.5, printed at the end of F4, and Strabo, for which see on F38; it is based on traders' tales (the chariots are similar), or stories going back to Caesar's invasions; 'flat' suits the areas Caesar reached.

F13 Mela 3.54–7

[54] triginta sunt Orcades angustis inter se diductae spatiis, septem Haemodae contra Germaniam vectae. in illo sinu quem Codanum diximus Scadinavia, quam adhuc Teutoni tenent, et ut fecunditate alias ita magnitudine antestat. [55] quae Sarmatis adversa sunt ob alternos accessus recursusque pelagi, et quod spatia quis distant modo operiuntur undis modo nuda sunt, alias insulae videntur alias una et continens terra. [56] in his esse Oeonas, qui ovis avium palustrium et avenis tantum alantur, esse equinis pedibus Hippopodas et Panotios, quibus magnae aures et ad ambiendum corpus omne patulae nudis alioquin pro veste sint, praeterquam quod fabulis traditur, apud auctores etiam quos sequi non pigeat invenio. [57] Thyle Belcarum litori adposita est, Grais et nostris celebrata carminibus. in ea, quod ibi sol longe occasurus exsurgit, breves utique noctes sunt, sed per hiemem sicut aliubi obscurae, aestate lucidae, quod per id tempus iam se altius evehens,

---

4 This is a lovely instance of how accurate information gets misunderstood and mangled in oral transmission. Rivers flowing back on themselves (*retro fluentia*) are mud flats or tidal estuaries; some had oyster-beds. But edible oysters produce only tiny, gritty pearls. It is river mussels which produce quality pearls. 'Gems' are an imaginative addition, unless amethysts and other semi-precious quartz stones that are found as pebbles in rivers and on beaches in Scotland were being traded to the south of England; Pliny, when dealing with gems in book 37, does not know of Britain, much less Scotland, as a source (e.g. *NH* 37.32, 40).

quamquam ipse non cernatur, vicino tamen splendore proxima inlustrat, per sol-stitium vero nullae, quod tum iam manifestior non fulgorem modo sed sui quoque partem maximam ostentat.

[54] *There are thirty Orcades, separated by narrow spaces between them. There are seven Haemodes extending opposite Germany. In what we have called Coda-nus Bay there is Scadinavia, which the Teutoni still hold, and it stands out for its fertility as well as its size.* [55] *The islands opposite the Sarmatians, on account of the ebb and flow of the tide, and because the spaces between them are sometimes covered in waves and other times bare, seem to be islands at some times and other times land.* [56] *Among these are the Oeonae, who feed mainly on the eggs of marsh birds and oats, the Hippopodes, with horses' feet, and the Panotii, who have large ears spreading round their whole body for clothing: they are otherwise naked, as is handed down in legend and in authors whom I find there is no shame in following.* [57] *Thule, celebrated in Greek poetry as well as our own, is situated near the coast of the Belcae. On it, because the sun rises a long time before it will set, the nights are therefore short. In winter they are as dark as anywhere, and in summer, bright. This is because at that time the sun goes higher in the sky, and even when it is not seen, it lights up the adjacent parts with its nearby brightness. But at the solstice there are no nights, because then it is even more visible and shows not only its radiance but the greatest part of itself also.*

After describing Britain, Mela turns to the islands around it. As noted on F12, he first mentions Ireland, located above what he calls the 'rear side', the west coast, of Britain.[1] Our text follows. Two details need noting: (1) it shows no knowledge of Jutland as a large peninsula, and (2) Mela's location of the Orcades and Thule seems wrong; it suggests that his knowledge of this part of northern Europe was vague. Map 2 attempts to show how he apparently envisaged the whole area (cf F12).

A few pages earlier, he has described the Ocean coast of Germany and places further east. Germany begins at the Rhine and extends to Sarmatia, whose coast faces north and the Ocean (3.25). Beyond the Albis, the Elbe, there is the huge Codanus Bay, full of large and small islands. For this reason, the sea does extend very widely, but is spread out like rivers, hemmed in by the islands near to each other, like a strait. On the bay are the Cimbri and Teutoni; further on, the Herm-iones, the last German tribe (3.31–2). After Germany there is Sarmatia, which extends to the Vistula; Scythia begins after that. Most of the Scythians are called Belcae (3.33, 36). Given Mela's perception of the area, it is impractical to put his description on a modern map.

§54 The Orcades derive from Pytheas; they, and their number, are discussed on F19. However, there is a problem as to where Mela locates the Haemodae. It

---

1 For Mela, 3.53, the inhabitants are 'ignorant of virtues' and 'totally lacking in piety'; for Strabo's similar location see on F25 and Appendix 7; its inhabitants, p 8 n 5.

is both textual and substantive. The latter is noted in relation to §§55–6. In the MS, the words 'ex iis' ('among them') were written before 'Scadinavia', and then crossed out. They are omitted here, with a full stop after the Haemodae reference, so treating them as distinct from Scadinavia. Silberman (1988) retains them, with a comma after the Haemodae reference, so placing them in the Codanus Bay next to Scadinavia.[2] On any view, given Mela's belief that the apex of Britain pointed north-east, if he believed that the Orcades lay east of the apex and the Haemodae further east still, his location of the latter somewhere opposite Germany ('vectae contra', 'carried opposite') was logical. The problem of actually locating the Haemodae, and its spelling, are discussed on F19. Mela differs from Pliny in naming the inhabitants of Scadinavia as Teutoni.[3] It should be noted that Mela's MS spells the island 'Codannovia', universally amended to 'Scadinavia' to correspond to the similarly spelled name in Pliny, F17, Scatinavia.[4] As there, it refers to southern Sweden, thought to be an island. Mela offers the first mention of both Sweden and the Vistula.[5] Although he covers some of the same ground as Pliny in F17, his account appears to come from fewer and older sources.

§§55–6 The islands in these paragraphs are problematic. The wording is not dissimilar to that of the large and small islands in the Codanus Bay in 3.31 noted above; both are garbled accounts of islands affected by tides. But Mela presents those here as a separate group, and includes the three islands of §56. For the latter, he must have used similar sources to Pliny. Pliny understood them to be vaguely in the north, and, as noted on F17, they probably go back to stories of peoples in northern Sweden or Finland; Mela makes them part of his islands off the coast of Sarmatia. The difficulty is that there are no islands off the coast of Sarmatia, i.e. Poland, west of the Vistula. Factually, there is only one group of islands in the Baltic: those which make up Denmark. Further east, there are no groups of islands which could answer Mela's description here or in 3.31. There are only Rügen and Bornholm. From Rügen eastwards, there are no islands, much less groups. However, the sea-ways around the Danish islands and up to the Fasterbro area of Skåne, Sweden were and are full of dangerous sandbanks and shoals: see Svennung (1963) 27–32, where the subheading is 'Schadeninseln', Dangerous Islands. But whether this corresponds to Mela's description of either of his groups of islands is a question.

2 Frick (1888) 68 prints the full stop after the Haemodae reference, but also amends 'ex iis' to 'eximia', 'excellent' as an adjective for Scadinavia. A 1711 editor, Reinoldius, suggested 'ex insulis' for 'ex iis'; 'among the islands' (i.e. the Haemodes) was Scadinavia. For Silberman's map, see n 7.
3 Mela knows that the Teutons also live on the Codanus Bay (Note to Appendix 4 §§4, 9). Pliny's Scadinavia inhabitants are the Hillevones, F17; a hundred years later, Ptolemy would inhabit it with Goths (Note to Appendix 4 n 18).
4 Mela survives in only one relevant MS, Vatican library, no 4929; all other MSS are copies of that. There are variants in Pliny's MSS, but 'Scatinavia', if not his spelling, is very close to it.
5 As Mela was used by Pliny for his book 4, it is probable that Mela was an older contemporary.

There are three ways of resolving this. One is that Mela's sources while in fact reflecting reports of Denmark described them somewhat differently, and misled him into thinking that there were two groups of such islands, one in the Codanus Bay, one off Sarmatia. Another is that one or more of his sources referred to the Frisian islands; though west of the Elbe, in an oral world without maps Mela could have thought that they were further east. A third is that in ancient times the Vistula reached the Baltic via a delta; the modern farmland is from alluviation. Mela's belief in a separate group could have a factual basis if his sources reflected reports of small tidal islands in the delta, or periodic flooding in the delta creating islands.[6] Any of these possibilities is shown on map 2. Note that I spell the tribe on the third island of §56 'Panotios' to correspond to Pliny; Mela's MS has the otherwise unknown 'Sannalos', a good example of an unfamiliar name getting mangled in transmission.[7]

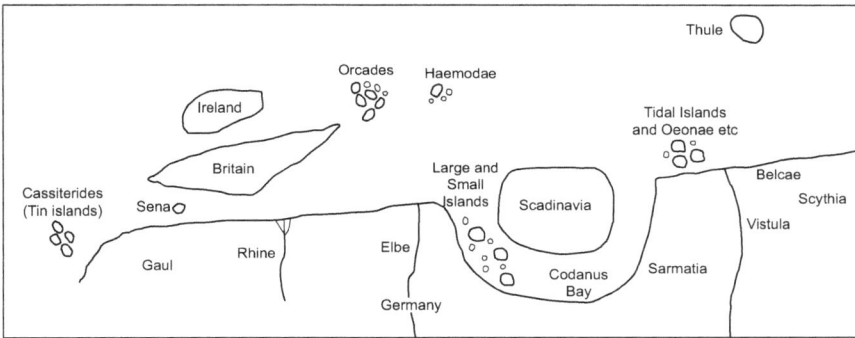

Map 2 Sketch map showing Mela's apparent beliefs as to northern Europe. It is unclear what relative dimensions he envisaged for Britain, Gaul, and Germany, or the distances between the Rhine and the Elbe, and the Elbe and the Vistula. All island locations must be treated as approximate.

Source: Drawn by the author.

6 See Silberman (1988) 287; map clearly showing the delta, Dion (1977) 214. Even today, the land each side of the main Vistula channel are technically two islands, as is clear from Google Maps, the eastern larger than the western. East of the Rostock area, there are no other coasts where there are offshore islands; Rügen, especially with lower sea levels, does not fit. Dion's map is marked as in the time of Pytheas; but the possible identification of the islands here with the Vistula delta does not prove that Pytheas got so far east.
7 The map with Silberman (1988) places the Haemodae just west of Scadinavia, where map 1 has Mela's large and small islands. He does not show the latter group, or the three islands of §56, on it. He shapes the west side of the Codanus Bay, with the Ocean coast of Germany running in a north-easterly direction up to it, so as to create a semblance of Jutland, which goes beyond what the picture in Mela's text suggests. He shapes the east side of the Bay so as to place Thule north of Scadinavia.

A separate point on these tidal islands, either here or in 3.31, is whether they preserve some of Pytheas' description of the area, summarised by Pliny in F21 as an 'aestuarium', a tidal coast: see Appendix 4 §§5–8, especially §8. Strictly speaking, they are distinct: Pliny speaks of a tidal coast, and Mela of tidal islands. It is tempting to propose that if we had Pytheas' full text, we could see that he did mention both islands and coast; on the evidence we have, we can only speculate, though the coincidence of a noteworthy tidal effect in an otherwise little-known part of the world is striking.

§57 No references to Thule in Greek poetry survive, but Mela will have known of 'ultima Thule', Virgil *Georg* 1.30, and perhaps Seneca *Medea* 379.[8] Indeed, Mela's location of Thule shows that he ignored Pytheas' location of it as north of Britain (F15, F24; cf p 3). That it was 'near the coast of the Belcae' makes sense only on Mela's own terms. Since he calls most Scythians 'Belcae' (see above), he thinks of Thule as opposite Scythia. But 'Belcae' is not a name recorded elsewhere, as discussed in Appendix 6 §10. One possibility is that it stems from a mispronunciation of the Bergi (F19), and then confusion as to where they lived. Mela's description of long summer nights and long winter days, and full daylight at the solstice, is accurate, and will go back to Pytheas. Unfortunately, what he says is too general to help resolve the question what Pytheas wrote about daylight on Thule: see Appendix 2 §8. For the spelling 'Thyle', see on F19.

F14 Pliny *NH* 1.2, 1.4, 1.37

(a) Pliny *NH* 1.2: Libro II continentur . . . ubi longissimi dies, ubi brevissimi . . . qua ratione aestus maris accedant et recedant, ubi aestus extra rationem idem faciunt . . . Ex auctoribus . . . Externis . . . Pythea . . .

Book 2 contents . . . where the days are longest, where shortest . . . why tides of the sea rise and fall; where extraordinary tides occur . . . From foreign authors . . . Pytheas . . .

(b) 1.4 Libro IV continentur situs, gentes, maria, oppida . . . populi qui sint aut qui fuerunt . . . insularum Ponti, Germaniae, insularum in Galico oceano XCVI, quas inter Britannia . . . Ex auctoribus . . . Externis . . . Timaeos Siculo . . . Pythea . . .

Book 4 contents: sites, races, seas, towns . . . populations present and past . . . of the islands of the Black Sea, of Germany, of the 96 islands of the Gallic Ocean, among them Britain . . . From foreign authors . . . Timaios of Sicily . . . Pytheas . . .

---

8 If *Medea* had already been published and also had come to Mela's notice.

(c) 1.37 Libro XXXVII continentur . . . de sucino: quae mentiti sint auctores de eo, genera eius VII . . . Ex auctoribus . . . Externis . . . Pit<h>ea, Tim<a>eo Siculo . . .

Book 37 contents . . . concerning amber: the lies authors have told about it, seven types of it . . . from foreign sources . . . Pytheas, Timaios of Sicily . . .

Pliny the elder, born 23 AD, had a distinguished military career. In 79 AD, he was *praefectus classis*, 'chief of the fleet', i.e. admiral, at Misenum, and died, probably from inhaling fumes from the eruption of Vesuvius, trying to rescue refugees stranded on a nearby beach. We are fortunate in having his biographical details in a letter by his nephew Pliny the younger. It lists his massive literary output at *Ep* 3.5.3–6, lost except for his 37 volume *Natural History*, a compendium or encyclopaedia encapsulating a vast quantity of information about the natural world. The letter mentions some of his military posts and adds that he also practiced law at one stage. It stresses that he read and made notes every minute of the day when not working, and when travelling dictated to a shorthand writer ('notarius', *ibid* 3.5.15; generally 3.5.7–17).[1]

The above texts are from book 1, which combines a list of contents and a bibliography. The latter is divided between Latin and Greek ('externi', *foreign*), and taken together they reflect a vast quantity of reading. A typical list has a dozen or more authors in each language for each book. Unlike Strabo, he had no difficulty in accepting what Pytheas had written. For northern Europe, he could use two writers not available to Strabo: Mela, for whom see on F12, and Philemon (p xx).

The implications of what his nephew wrote suggest two things: that he always had secretaries to make notes on and summaries of his sources; and that when he wrote his actual work, he compiled it from his and their notes rather than the actual book. This is assumed here, and helps to explain discrepancies, for instance the same island(s) with different names in the Baltic. We may infer that his notes for his geography sections, drawn on for F17 from book 4, were a different set from his notes about semi-precious stones such as amber, used for F21 from book 37. When writing the latter, it would not necessarily occur to him to cross-check what he had written about the geography of the Baltic some time, perhaps a couple of years, earlier.[2] Also, finding one passage in a papyrus roll is much harder than in a printed book. There was also the risk of errors from the recording and later transcription of unfamiliar foreign names by his secretaries, as well as the perils of later MS transmission.

1 For both career and total literary output, see *OCD* and *BNP* sv.
2 When he could get home to Lake Como, he would have copies of what he had previously written. If he was writing when at a military posting, he might not have had access to all of it. Errors of one sort or another can be found elsewhere in his work, e.g. *NH* 4.18, the Nemean Games being held every two, not four, years.

F15 Pliny *NH* 2.186–7

[186] sic fit, ut vario lucis incremento in Meroe longissimus dies XII horas aequinoctiales et octo partes unius horae colligat, Alexandriae vero XIIII horas, in Italia XV, in Britannia XVII, ubi aestate lucidae noctes haud dubitare repromittunt,
id quod cogit ratio credi, solstiti diebus accedente sole propius verticem mundi
angusto lucis ambitu subiecta terrae continuos dies habere senis mensibus noctesque e diverso ad brumam remoto. [187] quod fieri in insula Thyle Pytheas
Massiliensis scribit, sex dierum navigatione in septentrionem a Britannia distante,
quidam vero et in Mona, quae distat a Camaloduno Britanniae oppido circiter CC,
adfirmant . . .

[186] *So it happens that with the varying increase in light the longest day at
Meroë amounts to 12 equinoctial hours plus eight parts (twelfths) of an hour, at
Alexandria 14, in Italy 15, in Britain 17, where the light nights in summer assure
(us) without doubt that which reason compels us to believe: during the days of
summer, when the sun comes nearer to the top of the world, with a narrow circuit
of light, the parts of the earth below have continuous days for six months, and
[continuous] nights when [the sun] is removed in the opposite direction at the
winter solstice. [187] That this is so in the island of Thyle, six days voyage distant
to the north of Britain, Pytheas of Massilia wrote. Indeed some also assert this on
Mona, which is distant from the British town of Camalodunum [Colchester] by
about 200 miles.*

Pliny's book 2 is an overview of the world; books 3–6, from which F17 to F20
come, deal in detail with specific areas. This text lies within a section discussing
the point that day and night are not the same everywhere.

§186 For 'equinoctial hours', see on F7. The daylight lengths are not quite the
same as those in Cleomedes II 1 438–44 referred to on F2, or Hipparchos, Appendix 2 §§5–6 and on F28; probably several such lists existed.[1] For the spelling of
Thule, see on F19. There is a textual problem with the MSS 'repromittunt'. The
sense is clear; I translate 'assure'.[2] The plural 'diebus' (days) shows that Pliny
uses 'solstitium' not in its literal sense of the solstice ('sol', sun, 'stit-', stand), but
the days of high summer. His 'narrow circuit (or circle) of light' seems to convey
the idea of the sun moving in an ellipse above the horizon and never setting. Pliny
has not fully grasped, or perhaps confused, the two components in the theory
hinted at in Cleomedes, F2, and explicit in Geminos immediately after F7, 6.10–
22: the further north you go, the longer the days; eventually there is continuous

---

1 Thus Hipparchos' figure for Meroë is 13 hours (Strabo 2.5.36).
2 'Repromittunt' literally means 'promise in return'. Beaujeu (1950) 82 translates: the light nights
'garantissent sans conteste' that which theory obliges (you) to believe, adopted here. Suggested emen-
dations are 'se promittunt' (Mayhoff); and 'permittunt' (Detlefsen); in the context, both would have
the connotation 'allow' rather than Beaujeu's 'guarantee'.

daylight for a month or more, and six months at the north pole. Pliny did not understand that it was only in the area of the north pole that daylight would last for six months.

§187 This is our only explicit reference to what Pytheas said about daylight on Thule. It is repeated without the attribution in F20. In F19, where Pliny uses Pytheas for island names, he speaks of the daylight more accurately by our standards. However, it is the only evidence that we have as to what Pytheas wrote: see Appendix 2 §8. The references to Mona, Anglesey, are puzzling. Any of Pliny's colleagues who had served in Britain would know that there is no continuous daylight in summer, though somewhat shorter nights than at home. Caesar, *BG* 5.13.4, says that 'some writers' say that there is a 30 day night on Mona in winter; stories about the island must have circulated. Anglesey was conquered in about 60–61 AD, and the Iceni destroyed the then capital, Colchester, about the same time. They are in fact over 300 miles, Roman or British, apart (say 480 km), and after 60 AD the actual distance would be known. If the number is accurately transmitted, it could reflect what was believed in Colchester before 60 AD. But whether the Romans were using Pytheas' or Caesar's dimensions for Britain (see on F4), the cross-country distance of 200 miles is too small. It is possible that an original 300, 'CCC', has been miscopied as 200, 'CC'.[3]

F16 Pliny *NH* 2.217

Omnes autem aestus in oceano maiora integunt spatia nudantque quam in reliquo mari, sive quia in totum universitate animosius quam parte est, sive quia magnitudo aperta sideris vim laxe grassantis efficacius sentit, eandem angustiis arcentibus. qua de causa nec lacus nec amnes similiter moventur. octogenis cubitis supra Britanniam intumescere aestus Pytheas Massiliensis auctor est.

However, all tides in the Ocean cover and uncover greater areas than in the rest of the sea, either because its movement is more lively in the universe as a whole than in a part, or because a large open space feels the power of the star [the moon] more powerfully as it moves over a wide area; that power being limited in narrow spaces. For this reason neither lakes nor rivers move in the same way. Pytheas of Marseille writes that above Britain, the tides swell to a height of 80 cubits [40 feet].

As noted on F15, Pliny's book 2 is an overview of the world; this text comes from §§212–220, a long exegesis about tides, starting with the influence of the sun and moon, especially the moon. Here, he contrasts the modest tides

---

3 The distance is about 235 miles, 375 km, as the crow flies; by modern roads over 300 miles, 480 km. Two hundred Roman miles is about 185 miles, 295 km (p xvi). Once conquered, 300 Roman miles (280 miles, 445 km) as a round figure would have been a realistic assessment.

in the Mediterranean with those in the Ocean, and says that lakes and rivers have even smaller tides. Unlike the many names in Aetios, noted on F1, Pliny only mentions one name in the whole section, Pytheas. 'Above Britain' suits the Pentland Firth, the sea between Caithness and the Orkneys, which is notorious for its dangers. Carpenter (1973) 172 cites the Admiralty *North Sea Pilot* as speaking of '60 foot waves and columns of spray above them when gales blow counter to the tidal race'; the Shetlands are also notorious for bad weather, where Pytheas will have overwintered (Coda p 174). Pliny's words are a summary or paraphrase of what Pytheas actually wrote, and we can read them either as Pytheas describing his experience of the Pentland Firth, or seeing huge waves beat against a rocky shore or cliffs in the Shetlands. In either case, 80 cubits, 40 yards, would be a realistic estimate; and if 'tides' does not come from Pliny's notes it could certainly be how Pytheas expressed that which he experienced or saw.

F17 Pliny *NH* 4.94–7

[94] Exeundum deinde est, ut extera Europae dicantur, transgressisque Ripaeos montes litus oceani septentrionalis in laeva, donec perveniatur Gadis, legendum. insulae complures sine nominibus eo situ traduntur, ex quibus ante Scythiam quae appellatur Baunonia unam abesse diei cursu, in quam veris tempore fluctibus electrum eiciatur, Timaeus prodidit. reliqua litora incerta signata fama septentrionalis oceani: Amalcium eum Hecataeus appellat a Parapaniso amne, qua Scythiam adluit, quod nomen eius gentis lingua significat congelatum. [95] Philemon Morimarusam a Cimbris vocari, hoc est mortuum mare, inde usque ad promunturium Rusbeas, ultra deinde Cronium. Xenophon Lampsacenus a litore Scytharum tridui navigatione insulam esse inmensae magnitudinis Balciam tradit, eandem Pytheas Basiliam nominat. feruntur et Oeonae, in quibus ovis avium et avenis incolae vivant, aliae, in quibus equinis pedibus homines nascantur, Hippopodes appellati, Panotiorum aliae, in quibus nuda alioqui corpora praegrandes ipsorum aures tota contegant. [96] Incipit deinde clarior aperiri fama ab gente Inguaeonum, quae est prima in Germania. mons Saevo ibi, inmensus nec Ripaeis iugis minor, inmanem ad Cimbrorum usque promunturium efficit sinum, qui Codanus vocatur, refertus insulis, quarum clarissima est Scatinavia, inconpertae magnitudinis, portionem tantum eius, quod notum sit, Hillevionum gente quingentis incolente pagis: quae alterum orbem terrarum eam appellant. nec minor est opinione Aeningia. [97] quidam haec habitari ad Vistlam usque fluvium a Sarmatis, Venedis, Sciris, Hirris tradunt, sinum Cylipenum vocari et in ostio eius insulam Latrim, mox alterum sinum Lagnum, conterminum Cimbris. promunturium Cimbrorum excurrens in maria longe paeninsulam efficit, quae Tastris appellatur. XXIII inde insulae Romanis armis cognitae . . .

[94] *We must now move on from there [the Black Sea] to speak of the outside of Europe. Crossing the Ripaean mountains we must go along it with the coast of the Ocean on our left until we reach Gades [Cadiz]. Many unnamed islands in that area are reported. Timaios reported one called Baunonia, a day's sail from Scythia, where amber is cast up by the waves in spring. The rest of the coasts of the northern ocean are marked by uncertain reports. Where it washes Scythia beyond the Parapanisus River, Hecataeus calls it Amalcius, because the name means 'frozen' in the local language. [95] Philemon says that it is called Morimarusa by the Cimbri, that is Dead Sea, from there as far as the Rusbeae promontory, and then beyond that Cronian. Xenophon of Lampsacus reports a very large island, Balcia, a three days' voyage from the coast of the Scythians. Pytheas calls this Basilia. There are also reports of islands called Oeonae, where the inhabitants live on birds' eggs and oats; others where the people are born with horses' feet, and are called Hippopodes; and others of the Panotii ['All-ears'] where their otherwise naked bodies are completely covered by huge ears. [96] From here the reports are clearer, beginning with the first of the German clans, the Inguaeones. Mt Saevo is there, large and no smaller than the Ripaean chain. It reaches to the huge Codanus Bay by the promontory of the Cimbri, which is full of islands, of which the most famous is Scatinavia, of unknown size. The part that is known is occupied by the Hillevones in 500 districts; they call it the second circle of the world. Nor is Aeningia thought to be any smaller. [97] Some report that these areas, right up to the river Vistula, are inhabited by Sarmati, Venedi, Sciri and Hirri, that there is a bay called Cylipenus, and an island in its middle, Latris, and then another bay, Lagnus, which goes to the frontier of the Cimbrian lands. The Cimbrian promontory projects a long way from the land and makes a peninsula, called Tatris. After that there are 23 islands [i.e. further west] known to Roman arms.*

After his overview of the world in book 2, Pliny deals with individual areas. Book 3 covers the western Mediterranean; book 4 starts with Greece and the Balkans, and then the Black Sea, which he has just completed. He now moves to the Ocean coast of Europe, starting with the area east of Jutland. This section is printed in full not only for the context of the Pytheas reference, but also because it helps when looking at the other Pytheas references to amber and an island, as discussed in Appendix 4. After it, Pliny moves on to Germany west of Jutland with an overview of Germany, and then turns to Britain in F18. He will then continue down to Gades, the Latin form of the Greek Gadeira, Cadiz.

The whole must be read on the footing that, apart from Pytheas, we only know of one other eyewitness to the area: the expedition round Jutland in 5 AD referred to below. The rest will derive from authors noting reports circulating orally, basically from what merchants and traders said, bringing goods from the north over the last stages of their journey to the Adriatic and Mediterranean. Perhaps Pliny also had tales about the north when he served in Germany, or from colleagues who had served there. Unlike Mela, F13, the Jutland expedition gave him a good idea of the general shape of the coast, with bays east and west of Jutland; but, like others, he had no idea that the Baltic was a cul-de-sac, ending in the Gulfs of Bothnia and Finland. For him, it was the Ocean which washed the coast east of Jutland, and any land further north was an island; we should be alive to the possibility that some of what he reports referred to (say) Latvia or Estonia.[1]

§94 The Ripaean mountains have been noted at p 11; it is odd that Pliny names them here, as he usually speaks of them as being well to the north of the Black Sea and the Caucasus (e.g. *NH* 4.78, 88; 5.98; 6.15; cf on F39). Here they would correspond to the Carpathians in central Europe.[2] Timaios' island Baunonia and Xenophon's and Pytheas' Balcia/Basilia islands are discussed in Appendix 4. Hecataios is Hecataios of Abdera, c 360–290; the details are from his *On the Hyperboreans*, a work of fiction.[3] He is of particular interest because his book may well be a parody of Pytheas; if so, it would help confirm Pytheas' dates (pp 166–7). His Hyperboreans live on a distant northern island; you pass the Amalcius coast and Parapanisus river on your way there.[4] But neither name is 'foreign'. Amalcius is more or less a Greek

1 Pliny's reporting anything he can extract from his sources may be contrasted with the scientific approach of Ptolemy, a hundred years later (*Geog* 2.11.16), who notes only places for which he could offer co-ordinates. In the Baltic, Ptolemy only has three islands, 'Alocia' above Jutland, and four islands ('Scandiae'), one much larger than the others called 'Scandia'. Denmark consists of many islands; Læsø possibly derives from Alocia, and other Danish islands are presumably partly reflected in the Scandiae, with the large Scandia being Pliny's Scatanavia of §96. On a detail, note that the north of Jutland only became an island in 1825, when the North Sea broke through and gave the Limfjord exits to both it and the Kattegat. Ptolemy's co-ordinates are wrong by our standards, and so do not help on identification.
2 Only in Ptolemy (*Geog* 3.5, 7.1, 8.1) do we find the actual Carpathians so called (Καρπάτης, *Karpatēs*).
3 Pliny treated this work as factual, presumably because Hecataios also wrote factually about Egypt, where he lived for some years around 300 AD: Pliny cites that work in book 5 within his geography of Egypt, and in book 18 on corn in Egypt.
4 The fragments are at *FGrH* 264 F7–14; there is an English translation by Raffael Joorde (2016). The island is called Elixoia, 'no smaller than Sicily' (cf on F4); it is 'fertile with many crops, with a mild climate producing two harvests per year' (so Diodoros 2.47.1, attributing this to 'Hecataios and others'. Their way of life, religion, and language, are described with a good deal of imagination. The crops and the frozen sea references could both have been inspired by Pytheas (crops, cf F38; frozen sea, Appendix 6 §§14–16). Hawkes (1967) 38–9 argues that Hecataios' round temple to Apollo on Elixoia reflects a description of Stonehenge in Pytheas, which assumes both extensive inland travelling by him and that he recorded it in his book. But Apollo's round *tholos* at Delphi is a more likely inspiration, and Bridgman (2005) 137 questions Hawkes' argument.

word, meaning either 'not cold' or 'very cold'.[5] It cannot be the actual local name of any part of the Baltic, and is not evidence that Hecataios knew that the northern Baltic can freeze; at most he might know that the sea in the Crimea can freeze, as noted by Herodotos (4.28.1; cf p 159 n 18). It was long believed that the Caspian had an outlet into the northern Ocean; Parapanisus is like 'Parapamisos', the name known since Alexander's conquests for the Hindu Kush. It is easy to envisage Hecataios hearing it with an -n- in the middle and using it as a suitable name for a river imagined as rising in India, flowing into the Caspian Sea, and thence to the northern Ocean.[6]

§95 The Greek geographer Philemon is noted on p xx; his 'Dead Sea' presumably goes back to reports that parts of the Baltic can freeze.[7] The name, 'Morimarusa', is close to *mors*, death, and *mare*, sea, as if a parody-word to convey the idea of 'dead sea' to a Latin speaker, a *mor-mare*.[8] The Cimbri lived in Jutland (cf §96 below), so Philemon's source knew something of that which Pytheas heard on the spot, that in spring the waves are stronger from melt-water (F21). 'Cronian' as an adjective for a distant northern sea is repeated in F19, where is it noted. Xenophon of Lampsacos, c 100, wrote a *Periplus*, based on oral reports.[9] His island Balcia, and Pytheas' Basilia, are discussed in Appendix 4. The next three islands are classic examples of stories which could have a kernel of truth but improve in the telling. The diet on the Oeonae islands recalls that on remote Scottish islands until recent times, and still the case in the Faroes. The fact that the practical and military Pliny saw fit to record stories of men with horses' feet, and huge ears for clothes, is a matter of comment. But they could have started from stories, say of the Sami, of snow shoes or boots with small skis, and one-piece hooded fur garments, originating in northern Sweden or Finland.[10]

---

5 μάκλιος, *malkios*, 'freezing', 'benumbing'. The a- at the beginning would normally imply a negative (so the 12th century AD *Etymologicum Magnum* has ἀμακλεῖν, *amalkein*, 'not to be cold'); but can sometimes be intensive, so 'very cold'.

6 So Mela 3.38, within his description of Scythia (just after text (iv) in the Note to Appendix 4). The idea may have originated with Alexander the Great; generally Romer (1998) 111 n 17.

7 The author found frozen sea water as far south as a beach in Latvia in April 1994, although the wider sea was open. But further north, the sea can freeze.

8 At first blush, a genuine northern origin for the word could be claimed, since *meri* means 'sea' in Finnish and Estonian; but it would be a false friend, as the word entered those vocabularies from a Germanic or Baltic source. Bridgman (2005) 129 wrongly attributes the name to Pytheas, as well as calling it a Celtic name; cf p 11 n 15.

9 Pliny *NH* 6.200 cites him for the Gorgons, a tribe of Amazonian women warriors, extinct long before the Trojan war, who lived on an island off Africa at the edge of the world; their story is told at length in Diodoros 3.52–55; and for kings in Illyria who lived for 600 and 800 years, *NH* 7.155.

10 Hdt 4.25.1 notes a tribe in the north of Scythia reporting that men with goats' feet live even further north, probably also a garbled account of snow shoes. Solinus, c 400 AD, in his *Collectanea rerum memorabilium*, *Collection of Noteworthy Things*, repeats the substance of Pliny, but ascribes all three islands to Xenophon (19.6). His book is largely, but not entirely, a reworking of Pliny. There is a MS problem about the name of the huge eared folk. I print 'Panotii' following Isidoros of Seville, c 600 AD, in his *Origines*, 11.3.19, though that might be his rationalisation, as the word connotes 'all-ears' in Greek. Pliny's MSS offer Phaenesi-, Fanesi-, and Fanesti-orum. Mela, F13, presumably from the same source(s) as Pliny, has the same islands spelled 'Sannalos'.

§96 The reference to the first tribe in Germany suggests that Pliny wants to move from east to west; but some of the names that follow, §97, may relate to places in the eastern part of the Baltic; it is doubtful how far he could have drawn a sketch map showing the relative positions of all the places he mentions. Also, by our standards, his confidence in his 'clearer sources' was misplaced. That the Inguaeones are the most easterly German clan is inconsistent with what he writes a few pages further on, *NH* 4.99–100, where he has the Vandals living between the Inguaeones and the Vistula, and so the first German tribe (text (vii) in the Note to Appendix 4, where it is discussed).

The Codanus Bay is clearly part of the Baltic, and it is easy for us to think of it as between Jutland and the Vistula; but it has to compete with the two bays in §97, and since it occurs in Mela, F13, all we can say with certainty is that it was one of the names that circulated that referred to the Baltic or some part of it. For the 'huge' Mt Saevo, Pliny's sources failed him. There are no mountains near the Baltic; a very substantial area of Germany both west and east of Jutland is called the North German Plain. In so far as Mt Saevo had a factual basis, we could think of reports of the Alps, or, further east, the Sudetes, the boundary between the Czech Republic and Poland, or the Tatras further east still, essentially the western end of the Carpathians, misreported as stretching much further north than they do; the foothills of these ranges are at least 350 miles, 560 km, from the coast.[11]

Since, as noted above, any land further north had to be an island in the Ocean, Scatinavia, Mela's Scadinavia, F13, is in fact southern Sweden. There were different stories about its inhabitants. Mela, F13, has the Teutoni. Pliny calls the Hillevones a *genus*, clan.[12] Tacitus records a German tribe, Helvaeconae, *Germ* 43. Pliny may be referring to a part who had migrated to Sweden: see Zehnacker and Silberman (2015) 322. Ptolemy *Geog* 2.11.16 has Goths and Phinnoi (Finns?).[13] The large Aeningia could be Sjælland, at the western end of the Baltic, but in view of what follows in §97 it is more

---

11 On the face of it, 'Saevo' is 'wild' or 'savage' (Latin 'saevus'). Perhaps it and the name stemmed from the report of a Roman official in Raetia or Noricum, experiencing the Alps, and believing that they extended (or must extend) a long way to the north. Despite how Pliny describes it, some propose southern Norway: see Zehnacker and Silberman (2015) 321.

12 Roman authors usually call a tribe just by their name; 'tribus' meant one of the tribes into which the Romans themselves were grouped. See text (vii) in Appendix 4, p 146. His '500' districts is probably an exaggeration in his sources.

13 Iordanes, c 550 AD, himself part Goth, and so likely to be familiar with their lore, wrote that Goths originated in 'Scandza' (*Get* III.16–IV.25: his Latin spelling would reflect his pronunciation). Tradition is not always accurate, but it is thought that a North Germanic dialect was already spoken in 'Scadinavia' in Pliny's time.

probably further east: Öland or Gotland, or some part of the Baltic States or even Finland.[14]

§97 Clearly, names circulated which originated in the Baltic but with little detail as to just where they were. If the 'areas . . . up to the Vistula' meant west of the river, Pliny accessed a similar source to Mela, F13. But Mela put the Sarmatians there; elsewhere in Pliny, the Sarmatians are always east of the Vistula and in Scythia: *NH* 2.246, 4.41, 4.49, 80–1, 91. This is also the case for both the Sciri and Venedi in other texts (Zehnacker and Silberman (2015) 322). The Hirri are otherwise unknown and could be a doublet for the Sciri, perhaps from a misread note. Zehnacker and Silberman, *loc cit*, suggest that Pliny's 'these areas' and the Sarmatians could be east of the river. He has just noted Aeningia. He did not know that the mainland coast turned north beyond the Vistula to become the coast of the Baltic states, and would believe that it continued as the coast of Scythia more or less to the east or north-east. But if he thought that Aeningia was an island north of that part of the Scythian coast, 'these areas' and the Sarmatians would be east of the Vistula. As to locating Cylipenus, Lastris, and Lagnus, we have to try and place them over a distance between 330 miles, 535 km, and 450 miles, 725 km.[15] Lagnus could refer to the area between the island of Rügen and Lübeck, or part of it, since the Cimbrian lands are Jutland, Tastris. On the other hand, Pliny might have thought that Cypelinus was in the Vistula area. If so, it and the island Lastris could reflect reports of the Bay of Gdansk, the Vistula Lagoon, and the strip of land separating them.[16] Latris has a Greek ending, but the context does not suggest that it was in Pytheas. On any view, it is not easy to see how the two bays differ from, or fit into, the Codanus Bay.

Pliny is the first author to give a clear description of Tastris, Jutland. In 5 AD, when the Romans still had some control up to the Elbe, a naval force under Tiberius sailed round Jutland (*RG* 26; Pliny *NH* 2.167; Vell Pat 2.104–8; with less detail Cass Dio 55.28). He probably used an official report;[17] at 2.167

---

14 The name is written *feningia* in some Pliny MSS, and this has led some to propose Estonia or Finland: Zehnacker and Silberman (2015) 322. Öland or Gotland are also possible identifications for Xenophon's Balcia: see Appendix 4. It appears that stories about the area beyond the Vistula were filtering through: at *NH* 2.246 Pliny says, 'I have information (cognitum habeo) of huge islands beyond Germany which have been recently discovered'. Perhaps these are the names here.

15 These are the distances as the crow flies from Lübeck to Gdansk and Kaliningrad respectively.

16 Alternatively, Bornholm and Rügen are possibilities; though the possible identification of Timaios' Baunonia with Bornholm (see Appendix 4 §2(c)) might tell against the first. Zehnacker and Silberman (2015) 322–3 canvass various suggestions for Latris on the basis that it is arguably at the west end of the Baltic, e.g. somewhere between Flensburg and Rostock.

17 In his list of authors for the book (*NH* 1.4) he includes 'the divine Augustus'; the report will have been published under the emperor's name. Our *Res Gestae Divi Augusti* ('Deeds of the Divine Augustus', hence *RG*) is a brief summary of his conquests and other achievements. *RG* 26 says that a fleet sailed from the mouth of the Rhine 'to the land of the Cimbri, to which no Roman had ever penetrated by land or sea'. Whatever Pytheas has said, as a mere Greek, it would not count.

he says that the expedition reached Scythia, 'a very wet area',[18] about which we can take two views. One is that, despite the more modest terminus in RG 26, n 17, it did reach at least to the Vistula. The other is that it reflects the rather loose geography of the Roman military, comparable to their mistaking the Shetlands for Thule, p 66, and they got no further than Jutland. It is clear from what follows after the end of our text that the 23 islands are west of Jutland, those off the west coast of Denmark and the Frisians, and perhaps including Helgoland.[19]

F18 Pliny *NH* 4.102

Ex adverso huius situs Britannia insula, clara Graecis nostrisque monimentis, inter septentrionem et occidentem iacet, Germaniae, Galliae, Hispaniae, multo maximis Europae partibus, magno intervallo adversa. Albion ipsi nomen fuit, cum Britanniae vocarentur omnes de quibus mox paulo dicemus. haec abest a Gesoriaco Morinorum gentis litore proximo traiectu L. circuitu patere XXXXVIII LXXV Pytheas et Isidorus tradunt, XXX prope iam annis notitiam eius Romanis armis non ultra vicinitatem silvae Calidoniae propagantibus.

Opposite this area is the island of Britannia [Britain], famous in Greek and our records; it lies between the north and the west, at a great distance opposite Germany, Gaul, Spain and most of Europe. Its own name was Albion, though all the islands of which we will speak below are called the Britannias. It is 50 miles away from the Gesoriac coast of the Morini tribe by the nearest crossing. According to Pytheas and Isidoros, its circumference is 4875 miles. It is now almost thirty years since our knowledge of it was increased by the Roman army, as far as the Caledonian Forest.

As noted on F17, Pliny has covered Germany and now moves on to Britain. At *NH* 4.101, he detailed the mouths of the Rhine and the tribes living there. It is the area 'opposite' to which he now turns. Like Mela, F12, and Strabo, F25, he thinks that Britain extends to the Rhine: see on F25 for a possible explanation. It is a matter of comment that although Pliny had information about Britain from colleagues who had served there, as we see from the reference to Anglesey in F15, he had not learnt that the Kent coast was not opposite the Rhine. Like Caesar, *BG* 5.13, he thinks of the western part in some sense opposite Spain. See, further, Appendix 7. Fifty Roman miles (46 miles, 74 km) from Boulogne (Gesoriacum in Latin, later Bononia) is a fair estimate: he is thinking not of the Folkestone or Dover area, but the Roman port of Rutupiae, Richborough, the main access to Britain in his time. It was the nearest practicable

---

18 'Scythicam ad plagam et umore nimio rigentia', 'to the Scythian shore and [places] numb with excessive moisture' (trans Rackham).
19 There are many attempts to identify all the places. A useful starting point is Bianchetti (1998) 201–4; cf Zehnacker and Silberman (2015) 316–23 and Roseman (1994) 83.

crossing, at any rate for the bulk movement of men and supplies for the Roman army in Britain.[1]

This text and Ps-Arist *De Mundo* 383b, of about the same date, are our earliest occurrences of 'Albion'; it is not in Strabo.[2] Pytheas called it Britain (strictly *Prettania*: see on F5), and that is how it was generally known. According to Rivet and Smith (1979), 'Albion' is Celtic, with the connotation of 'land' or 'our land', and 'Pritain' from an Indo-European root meaning 'cut', referring to tattooing or perhaps the use of woad. 'Albion' is known in Irish and Gaelic mediaeval usage up to the twelfth century AD, but otherwise only occurs in poetic or imaginative contexts.[3] Caesar knew only Britannia. This suggests that Albion was a local name picked up by the Roman military further north than Caesar reached, with some uncertainty as to how wide an area it meant. For the Romans, it remained Britannia; there is no evidence that after the Romans withdrew, it became 'Albion'; politically it remained Britain. As noted on F36 and the Coda p 174, Pytheas probably first learnt of Britain, and its name, at Corbilo on the Loire Estuary.

Isidoros of Charax in Babylonia, late first century BC–early first century AD, wrote mainly about his local area, but a book of geography with emphasis on distances was used by Pliny, who quotes all his distances in miles. For Europe, it must have been based on others' work, especially (it is thought) Eratosthenes. For Britain, Eratosthenes could only have used Pytheas, directly or perhaps from Timaios: 42,000 stades, F4. Thus Isidoros' figure was not independent of Pytheas, but Pliny would not appreciate that. Polybios' 'more than 40,000 stades' in F30 was from Pytheas' actual book, and confirms the accuracy of F4. It is not clear how Pliny derived his 4,875 miles; at the usual conversion of 8 stades to the mile (p xvi) it corresponds to 39,000 stades. It cannot be from Caesar's 2,000 miles (see on F4),[4] and we know of no one other than Pytheas who offered distances for all their coasts, much less actually sailed them. It is possible that there has been an error of arithmetic by one of his secretaries, or a mistranscription by a shorthand writer.[5]

---

1 It is thought that Claudius' then recent invasion of Britain used Boulogne. Caesar's invasions had used Portus Ictus, probably Sangatte or Wissant, between Boulogne and Calais.
2 For its occurrence in Avienus, see Appendix 1.
3 Rivet and Smith, *op cit*, 39–40; 248 (Albion), 280–1 (Britain). They showed that the once popular derivation from 'albus', white, and the White Cliffs of Dover connection, owed much to hope and imagination and is a false friend; the comparison with the mediaeval Welsh 'elfydd', 'land', from a root *Albio-, seems firm. Those who believe in the Massiliot Periplus can date some knowledge of Albion to Pytheas' time: but even on that basis, there is no reason that Pytheas had that knowledge: see Appendix 1.
4 Also, Caesar is not named as one of Pliny's sources for book 4 in *NH* 1.4.
5 Immediately after F18, Pliny continues: 'Agrippa believes that the length of [Britain] is 800 miles and its breadth 300 miles'. That at least shows that information of variable quality circulated. Agrippa, c 63–12, was an important military commander who wrote a *Geography* from which a map was put together after his death and publicly displayed; see *BNP* sv or *OCD* sv Vipsanius Agrippa.

The final sentence has some patriotic exaggeration. When Pliny died, it was only 26 years since Claudius' invasion in 43 AD, and the only Roman control of the north depended on the client realm of Brigantia and their queen Cartimandua. Agricola's attempt to bring Scotland into the empire was several years later. But the Romans had at least learnt that if you went far enough north there was an area called 'Caledonia', Scotland. Zehnacker and Silberman (2015) 332 suggest that the name is Greekoid ('consonance grecque') and originally from Pytheas. If so, we should be alert to the possibility, especially if he was sailing north up the west coast, that for him Caledonia began at the Firth of Clyde: Arran and then the Clyde estuary would suggest the natural border, as opposed to the modern political border, the Solway Firth (Coda p 181).

F19 Pliny *NH* 4.103–4

[103] . . . sunt autem XL Orcades, modicis inter se discretae spatiis, VII Haemodae, XXX Hebudes et inter Hiberniam ac Britanniam Mona, Monapia, Riginia, Vectis, Silumnus, Andros, infra vero Samnis et Axanthos et ab adversa in Germanicum mare sparsae Glaesiae, quas Electridas Graeci recentiores appellavere, quod ibi electrum nasceretur. [104] ultima omnium quae memorantur Tyle, in qua solstitio nullas esse noctes indicavimus, cancri signum sole transeunte, nullosque contra per brumam dies. hoc quidam senis mensibus continuis fieri arbitrantur. Timaeus historicus a Britannia introrsus sex dierum navigatione abesse dicit insulam Ictim, in qua candidum plumbum proveniat; ad eam Britannos vitilibus navigiis corio circumsutis navigare. sunt qui et alias prodant, Scandias, Dumnam, Bergos maximamque omnium Berricen, ex qua in Tylen navigetur. a Tyle unius diei navigatione mare concretum a nonnullis Cronium appellatur.

[103] *There are 40 Orcades, separated by moderate distances between them, 7 Haemodae, 30 Hebudes, and between Hibernia [Ireland] and Britannia Mona, Monapia, Riginia, Vectis, Silumnus, and Andros; Samnis and Axanthos are below; and, opposite the German Sea, the scattered Glaesiae, which the more recent Greeks have called the 'Electrides' because amber is produced there.* [104] *Last of all that are mentioned is Tyle, where we have noted that at the solstice there are no nights as the sun passes through the sign of Cancer; conversely, no days during the winter solstice. Some consider that this is true for six months at a time. The historian Timaios says that the island Ictis on which tin originates is distant from Britain inwards by six days sail; the Britons sail to it in boats made of hide sewn over with withies. People also mention other islands: the Scandiae, Dumna, the Bergi, and, largest of all, Berrice, from which one sails to Thule. From Thule by one day's voyage is the congealed sea, which some call Cronian.*

After Agrippa's dimensions for Britain (F18 n 5), Pliny gives Agrippa's dimensions for Ireland (600 by 300 miles), adding that the shortest crossing is 30 miles from where the Silures live (south Wales, so between Fishguard and Aberystwyth). Unlike Mela, F13, and Strabo, F31 and Appendix 7, who thought of Ireland as north of Britain, Pliny seems to know that it lies to the west. He now turns to the smaller islands: 'none are said to have a circumference of more than 125 miles', and continues as above. It is much more detailed than Strabo, whose description of Britain is noted on F38. It is clear that some names are from Pytheas, but Pliny had other sources. We cannot be sure how far he understood where each island or group was in relation to others, but it appears that he presented them broadly from north to south. The references to Anglesey in F15 and Claudius' invasion in F18 suggest that some of the names were from military colleagues who had served in in Britain. If the 200 miles in F15 is accurately transmitted, we could infer that book 2 was written around 60 AD, so that book 4 would be written a little later, about 20 years after Claudius' invasion. Britain was finally subdued in the decades after Pliny's death; the geographer Marinus could access fair detail around 110 AD.[1] There are some MS variations in the spelling of several of these names, but only one, 'Haemodae', is important here.

§103 The Orcades and Hebudes, with their Greek plurals, go back to Pytheas.[2] The Hebudes are the Hebrides, probably the Inner Hebrides, as in Ptolemy, *Geog* 2.2.10, *Alm* 2.6.28. Pytheas' spelling will have been Αἰβοῦδαι, *Aiboudai*, as in Ptolemy, but (H)Aebudae or (H)Ebudae when written in Latin.[3] The Orcades and Haemodae need considering together, as they raise several interconnected problems.

(1) The Orcades are clearly islands to the north of Cape Orca, F4, and are conventionally identified with the Orkneys, so Rivet and Smith (1979) 433–4. It is unclear how many Pytheas recorded. Mela, F13, has 30; Ptolemy *Geog* 2.3.14 has 'about thirty', and Isidore of Seville, c 600 AD, *Etym* 14.6.5, 33, perhaps turning 'about thirty' into a precise figure. Pliny's 40 is XL in the MSS; if from an error in transmission, it would be from XXX acquiring an extra X and then being rewritten as XL.

(2) 'Haemodae' is an editorial amendment in the light of Mela's Haemodae in F13. They are conventionally identified with the Shetlands, e.g. Rivet and Smith (1979) 241.[4] However spelled, they are otherwise unknown, and the question for us is not how later writers understood them, but what Pytheas recorded. Whatever

1 Marinus of Tyre was a geographer used extensively by Ptolemy, writing between 107 and 115 AD (*BNP* sv). He did not visit Britain but his work enabled Ptoleny to offer latitudes and longitudes for the British part of his *Geography*.
2 Note that the Orcades have a Celtic root and show that the inhabitants in Pytheas' time were Celts: Rivet and Smith (1979) 433–4. For the (H)ebudae, Inner Hebrides, *ibid* 354–5. In contrast to Pliny's 30, Ptolemy has six: a group of 5 and a further island with the same name: *ibid* 107, 145–6.
3 Pliny's 'H', silent in pronunciation, is decorative; by his time, there was little or no difference in the pronunciation of Greek *ai* and *e*: see Rivet and Smith (1979) 354–5, citing other evidence. In Ptolemy they are near Ireland: Rivet and Smith *op cit* 107, 131–2.
4 Pliny's MSS offer haecm-, hecm-, and acm-; Rivet and Smith (1979) 241, who identify them as the Shetlands, prefer 'Acmodae'.

we think of Mela's geography, Pytheas made them separate to the Orcades, and it is suggested that they are a doublet for the Hebudes, as raised by Rivet and Smith *loc cit*, noting that Ptolemy only had the Hebudes.[5] That raises the question whether Pytheas' 30 or 40 was intended by him to include the Shetlands.

(3) We must not be blinded by (a) Shetlands being separate to the Orkneys, at least once they were settled by Norsemen, and soon within the diocese of Bergen; nor (b) the mistake in c 80–82 AD by Agricola's naval officers, who glimpsed Fair Isle or perhaps the southern tip of Shetland Mainland, and reported that they had seen Thule (Tac *Agr* 10.6); they no doubt knew their Virgil (p 1). Tacitus' report gained traction in later literature;[6] there is also a late work of fiction in which a group of travellers reach Thule as part of their fabulous journey.[7]

(4) The relevant question is what Pytheas included in his 30 (or 40). Unless we assume that he meticulously sailed round and surveyed at least the Orkneys, the number is more consistent with his including the Shetlands. Autopsy as he sailed north from the mainland would indicate a string of islands stretching ever northwards; local information would supplement that, and he arrived at a realistic round number. There are about 10 principal islands in the Orkneys and 7 in the Shetlands, and a large number of small islands and islets.[8] Conversely if 30 or 40 was just for the Orkneys, and Pytheas used Haemodae for the Shetlands, it is surprising that they merited as few as seven. He would not see any political or social reason to distinguish between them. The absence of Shetlands in Ptolemy is significant. The counter-argument is that Pytheas would experience a good stretch of open sea beyond the Orkneys and past Fair Isle until he reached, say, Sumburgh, and would think it as a separate group.[9] He almost certainly overwintered in the Shetlands (Coda p 174), and would learn the local name. That the Haemodae (or Acmodae, n 4) are the Shetlands is deeply entrenched,[10] but if we dismiss the idea that Pytheas surveyed the Orkneys, the balance of probability is that Pytheas

---

5 How he came to place Thule at 63° N instead of the arctic circle, 66° N, is outside this commentary. There is an interesting paper by Dmitriy Shcheglov, 'Ptolemy's Latitude of Thule'; it is available on academia.edu. Even at 63° N it is well north of the Shetlands, which are around 60° 30′ N.

6 For Orcades as the Orkneys and Thule as the Shetlands, see Orosius (a Christian priest who wrote a world history, early fifth century AD) 1.76–9; Claudius Claudianus 31–2; cf Rutilius *De Reditu Suo* 1.499 (not 5.499 as Dion) (both c 400 AD); cf Dion (1966) 210.

7 The author, Antonius Diogenes, is usually, but not universally, dated to the second century AD on the basis of his work being known to Lucian, who wrote *True Stories*. The total itinerary involved, as well as more usual destinations, Hades, and one character reaching the sun and moon. There is a helpful Wikipedia entry on Diogenes.

8 Modern numbers depend on how many islets and skerries one includes, and which are inhabited. There would be fewer in an era of lower sea levels.

9 Depending on where you measure, it is some 55–60 miles, 90–95 km, from the north of the Orkneys to the south of the Shetlands.

10 Roseman (1994) 90 dismisses the problem in a few words: 'the Orkneys and Shetlands are securely identified'; cf Zehnacker and Silberman (2015) 333–4.

intended his round number to cover both groups, and Ptolemy followed him; and the Haemodae are a doublet for the Hebrides.

Mona is Anglesey (cf F15), and Monapia (probably) the Isle of Man. The latter has a range of spellings in our sources, but all are Greekoid, and one or both islands could have been first mentioned by Pytheas.[11] Riginia (or its variant spellings in Pliny's MSS) has a Latin form and cannot have come from Pytheas; its identity is uncertain.[12] Vectis is the Isle of Wight, and is not to be confused with 'Ictis' in F5; see further on 'Ictim' in §104, below. Silumnus is the Scilly Isles, which with lower sea levels was then basically one large island, hence the singular form.[13] Andros, a pure Greek name, cannot be located. There is a slight argument that, like 'Basilia' (F17 and Appendix 4), it was Pytheas' euphonious rendering of the Celtic name for (say) Bardsey Island or Lundy. As to those 'below' Britain, Samnis cannot be securely located,[14] but Axanthos is almost certainly Ushant, Pytheas' *Uxisamē* of F26.[15]

The Glaesiae are probably the Frisians; Pliny notes Glaesaria as one of the islands west of Jutland at *NH* 4.104 (just after F17), and also in connection with amber, no doubt washed up from the west coast of Jutland, at *NH* 37.42. As to the 'Electrides' name, as noted on F6, 'electron' was the Greek for amber, and the myth of Phaëthon led to the creation of an imaginary archipelago at the mouth of the Po called the Electrides.[16] A glance at Barrington 40 shows that, in a world of lower sea levels, there were effectively no islands there. But in the sailing manuals of Scylax 21.9 and Ps-Scymnos 374, they are islands on the *east* side of the Adriatic, off the coast of Croatia. As Greeks had long been sailing to the head of the Adriatic to buy amber,[17] it is possible that the 'Electrides' was as a nickname

---

11 Mona, Rivet and Smith (1979) 419–20. For the Isle of Man, *ibid* 410–11, where the various alternative spellings in our sources are discussed. They all begin with M- (Monaoida, Manavia, etc), and Monapia is a reasonable possibility; it is very hard to establish just what was the 'official' Roman spelling during their occupation of Britain. Caesar called it Mona (*BG* 5.13.3).

12 The other spellings are Rignia, Ricnia, Richea; it could be Ptolemy's Ricina, one of the Inner Hebrides (*Geog* 2.2.10 just noted). Some suggest, however, that it could be Rathlin (or Racklin, closer to the Irish Reachlain), off the coast of Northern Ireland; cf Rivet and Smith (1979) 355. That raises a question as to Pliny's source: it was off Pytheas' route; how could details of this headland have reached the Mediterranean? The same question may arise on Dumna, §104. Zehnacker and Silberman (2015) 334 identify it with 'Riduna' in the 3rd century AD Antonine *Itinerium Maritimum* (*Iter Mar*), Alderney, which ignores Pliny's 'between Ireland and Britain'. See also Rivet and Smith, *op cit*, 514.

13 For the Scillies, see Rivet and Smith (1979) sv 'Silina (?)', 457–9, where the different spellings in our sources, including in Pliny's MSS, are discussed; perhaps the 'Sicdelis' in the *Iter Mar*. Editions of Pliny usually print 'Silumnus'.

14 'Samnis' is a generally accepted emendation of variant spellings in the MSS. It has a Greek ending. An otherwise unknown British city 'Samnion' is recorded by Herod *De Pros Cath* 3.1.359 and Steph Byz sv. Alternatively it could be the 'Sarmia' of the *Iter Mar* (n 14), Guernsey.

15 See Zehnacker and Silberman (2015) 334–5 for some of the possibilities.

16 Apoll Rh IV.505–6, 580; Strabo 5.1.9.

17 See on F6 for this trade route for amber.

for a group of islands in the area where they used to put in *en route*; cf on F21.[18] It would then be easy to apply the (nick-)name to islands in the North Sea where amber was washed up.

§104 We have three Latin spellings for Θούλη, *Thoulē*. 'Tyle', only here, perhaps reflects one of Pliny's secretaries who could not pronounce 'th'. 'Thyle' and 'Thule' reflect variations in transcribing the Greek -ou- vowel into Latin.[19] Pliny has daylight at the solstice more correctly here, as compared to F15, but it is doubtful if it came from Pytheas: see Appendix 2 §8. For Thule, see Appendix 6, especially §§4–13. Timaios' island is problematic on several counts. The MSS have 'Mictim', otherwise unknown. I print 'Ictim' as proposed by Mayhoff.[20] While this island goes back to the passage in Timaios that Diodoros epitomised in F5, Pliny somehow got it wrong. His Latin is difficult grammatically and unrealistic economically and practically. He puts the island 'abesse', away or distant, from Britain, by six days sailing 'introrsus', an adverb meaning inwards or inwardly. It cannot mean 'under' or 'less than' six days; if the 'inwards' is correct, it cannot mean six days away from Britain, i.e. to France. It is nonsense to say that tin originated on an island six days off the mainland;[21] it came from the mainland (Cornwall), and Timaios said so, copying Pytheas, F5. Pytheas gave a clear account of how the tin is mined, smelted, and taken to the island just offshore in wagons. As there noted, it is St Michael's Mount, accessible at low tide. It is unrealistic to speak of moving tin to an island six days away in a small boat like a coracle, only a small quantity at a time. Some identify Ictis with Vectis, the eventual Roman name for the Isle of Wight; but it would take far more than six days to sail there from Cornwall in a coracle (and the same back home).[22] Much ingenuity has gone into explaining Pliny.[23] He did have a detail not epitomised by

---

18 Barrington 20 shows that all the islands in the Adriatic had their own names, but that would not prevent amber merchants having a nickname for one group. Pliny *NH* 3.152 and Mela 2.114 also include them among Croatian islands. However at *NH* 37.32, a page or so before F20, Pliny expressly says that they do not exist, in the context of being a source for amber; perhaps he overlooked the nickname possibility.

19 Thyle: here, F20, Pliny *NH* 2.246; Mela F13, Juv 15.112, Servius on Virgil Geor 1.30, and in later verse (Silius Italicus, Statius); Thule: Virgil *Geor* 1.30, Seneca *Medea* 379, Tac *Ag* 10.4.

20 Alternatively, 'Ictin' (Detlefsen), preserving its Greek ending.

21 The error here might stem from the tin islands, the Cassiterides, being at the back of Pliny's mind: he names them a few pages later, 4.119, and at 7.197.

22 As a rough guide, the road distance from Truro to Southampton is about 200 miles, 320 km, and a glance at a map shows that a coracle plying the adjacent coast would need to cover at least that sort of distance.

23 There is an enormous literature: see Bianchetti (1998) 145, 173. Hawkes (1978) 29–31 argued valiantly for the Ictis here being the Isle of Wight, in effect inventing a tin market there. But it is clear from F5 that tin reached France in the Lannion area, not Cherbourg. Zehnacker and Silberman (2015) 336 suggests that the six days has crept in here from a note about Thule being six days from Britain.

Diodoros in F5, that the locals have coracles. That must go back to Pytheas, who was there and noted them, perhaps as part of the local lifestyle.[24]

The 'other islands' which 'some report' are a curious collection. The Scandiae are known from Ptolemy (*Geog* 2.11.16), who placed three off Jutland, i.e. Denmark, and a larger one further east, the latter being the Scat(d)inavia of Mela F13 and Pliny F17, southern Sweden. Pliny, F17, knows that Scatinavia is east of Jutland, and it is odd that here he connects the similar Scandiae with Britain; see on F13 for Mela putting Scadinavia near the Elbe. Dumna is thought to be Harris and Lewis (Rivet and Smith (1979) 342, who note its Gaelic name of *Domon*), and quite likely from Pytheas; its being cited here and not with the Orkneys and the Hebrides could be explained by the way it was recorded in Pliny's notes. The two names Bergi and Berrice are unknown elsewhere, and are discussed in Appendix 6, §§10–11, concluding that the Bergi are Norwegian islands in the Bergen area, and Berrice either the Shetland island where he overwintered, or a settlement on the Bergi; and that both almost certainly come from Pytheas.[25]

As noted on p 3, the 'frozen sea' in the far north was already popular knowledge when Varro recorded it, a hundred years before Pliny. The topic has to be considered in the light of other references as well as the location of Thule, and is discussed in Appendix 6, §§14–16, where it is proposed that it refers to ice floes which Pytheas encountered in early spring after overwintering in the Northern Isles. As to the 'Cronian' name, Cronos was the father of Zeus, a figure from a very early stage of the world. It is easy to envisage some Greek reading or learning about the frozen sea in the distant north, a very strange phenomenon to any Mediterranean dweller, and thinking it as so primitive that it went back to a very early stage of the world, and choosing 'Cronian' as a suitable adjective. Philemon's source for the adjective in F17 would derive from that.[26]

---

24 Even if one believes in the Massiliot *Periplus*, Appendix 1, Timaios would not know that work, and got his coracles from Pytheas. Caesar had learned about them when he invaded Britain: he had some built in 48 to cross the Segre river in Spain (*Bel Civ* 1,54).

25 It has been suggested that Bergi, although a plural, is Hoy or South Ronaldsay (see Zehnacker and Silberman (2015) 336), but it is difficult to see how a name for one of the Orkneys could become known, and in any case Pliny is speaking of a group of islands.

26 'Cronos' was also the name for the planet Saturn. Apollonios had used the adjective for the Adriatic in Apoll Rh IV.327, but routing of Jason's *Argo* was geographically eccentric. It does, however, suggest that the idea of Pytheas' frozen sea being 'Cronian' circulated comparatively soon after his book became known. Our only other references to the Cronian sea in the far north are poetic and late: Dion Perieg 31–4 (second century AD: πόντον . . . πεπηγότα τε Κρόνιόν τε . . . νεκρόν, *ponton . . . pepēgota te nekron te . . . ponton*, 'sea [called] frozen and Cronian . . . dead; later still Orphica *Argonautica* 1081–2, κρόνιον . . . πόντον . . . νεκρήν τε θάλασσαν, *kronion . . . ponton . . . necrēn te thalassan*, 'Cronian sea and dead sea'.

F20 Pliny *NH* 6.219

Hactenus antiquorum exacta celebrevamus. sequentium diligentissimi quod superest terrarum supra tribus adsignavere segmentis, a Tanai per Maeotim lacum et Sarmatas usque Borysthenen atque ita per Dacos partemque Germaniae, Gallias oceani litora amplexi, quod esset horarum XVI, alterum per Hyperboreos et Britanniam horarum XVII, postremum Scythicum a Ripaeis iugis in Thylen, in quo dies continuarentur, ut diximus, noctesque per vices.

Up to this point we have set out the results of old [authors]. The most industrious of later [ones] have added three latitudes to the remainder of the earth: from the Tanais [Don] through lake Maeotis [the Sea of Azov], the Sarmatians as far as the Borysthenes [Dnieper], and then through the Dacians [i.e. Romania], a part of Germany, and the shores of the Gallic provinces which the Ocean embraces, where [the longest day] is 16 hours; a second through the Hyperboreans and Britain, 17 hours; the last a Scythian (parallel) from the Ripaean mountains to Thule, in which, as we have said, there are continual days and nights in turn.

Pliny covered Africa and Asia in books 5 and 6, and concludes with a summing up of world geography, here noting the longest days at different latitudes.[1] Prior to this text, Pliny sets out the lengths for seven latitudes further south, including the *gnomon* readings, for which see Appendix 2 §3. The most northerly runs through Bologna and Milan, on to Vienne south of Lyon, and then through the Pyrenees (not very accurate by our standards), with daylight of $15\frac{3}{5}$ hours. Although he says they are from older writers, they do not closely correspond to Hipparchos' latitudes, as to which see Appendix 2 §§5–6 and on F28; they rely on later sources who routed the latitudes through places now under Roman rule. We might note that Pliny's latitudes were not followed by Cleomedes; as noted on F2, his latitudes and daylight figures broadly reflect Hipparchos.[2] The latitude expressed as through Britain and the Hyperboreans has led to the imaginative but unrealistic suggestion by Bridgman (2005) 129 that Pytheas mentioned the Hyperboreans: it is difficult to see in what context he could have joined the two names. Pliny here repeats six months daylight on Thule, attributed to Pytheas in F15; this confirms that the text of F15 has been correctly transmitted. For what Pytheas said, see Appendix 2 §8. Pliny believed in Hyperboreans and the Ripeans: see

---

1 This text is omitted by Roseman (1994) and Bianchetti (1998); Mette (1952) included it, and indeed also printed the following section §220, relating to Egypt.

2 A full discussion is outside this commentary, but it is likely that Pliny's presentation reflects the works of two contemporaries of Cicero, Serapion of Antioch (not to be confused with others of the same name) and Nigidius, for whom see *BNP* svv. Serapion is not named in the index to book 6, though he is for books 4 and 5, and also book 2, where he is called *gnomonicus*, someone making or noting *gnomon* readings (*NH* 1.4, 1.5, 1.2). Nigidius is named for book 6 (*NH* 1.6). An additional problem is that although solstice *gnomon* readings should correspond to the lengths of day and night (Appendix 2 §3), Pliny's figures do not quite do so.

on F39. It is typical of his piecemeal method of working that he does not realise that he has already located the Ripaeans much further south, F17.

F21 Pliny *NH* 37.35–6

[35] Sotacus credidit in Britannia petris effluere, quas electridas vocavit, Pytheas Guionibus, Germaniae genti, accoli aestuarium oceani Metuonidis nomine spatio stadiorum sex milium; ab hoc diei navigatione abesse insulam Abalum; illo per ver fluctibus advehi et esse concreti maris purgamentum; incolas pro ligno ad ignem uti eo proximisque Teutonis vendere. [36] huic et Timaeus credidit, sed insulam Basiliam vocavit . . .

[35] *Sotacos thought that [amber] flows from rocks in Britannia, which he called 'electrides'; Pytheas, that there is a tidal coast occupied by the Guiones, a Germanic people, called Metuonis, 6,000 stades [680 miles, 1,090 km] long. From here, the island Abalus is a day's voyage away. It [amber] is carried to it in spring by the waves; it is a discharge from the frozen sea. The inhabitants there use it for wood on the fire and sell it to their neighbours, the Teutoni. [36] Timaios also thought this, but called the island Basilia.*

The last book of Pliny's *Natural History* deals with precious and semi-precious stones.[1] He covers amber at some length at 37.31–51. Greeks and Romans could not know that it is the fossilised sap of pine trees, but did know that in Europe it came from somewhere in the north; in fact from areas where the marine sands are rich in strata containing it: the west coast of Jutland and Samland, as noted on F6. But they had a marvellous collection of explanations, which Pliny meticulously records; they range from Greek myths, including the Phaëthon-Eridanos story noted on F6, and fantasies such as its being the urine of a lynx, males producing a redder and females a whiter type, to more factual explanations as here.[2] He came close to stating its correct origin at 37.42–3, that it is the resinous secretion of pine trees; this was a lucky inference, based on it smelling like pine when rubbed. Apart from the first sentence, the rest of §35 must be from Pytheas; only someone on the spot could learn that it is wave-borne.[3]

---

1 It also includes items such as (h)ammochrysos, perhaps quartz, and chalazias, perhaps flint or mica, 37.188–9.
2 For instance, just before the above text he cites Philemon, that amber comes from two places in Scythia, white from the one, called 'electrum', and red called 'sualiternicum' from the other (*NH* 37.33, just before the lynx urine story). He reports one Nicias, that tides cast it up on the shores of Germany (factually correct, but with a fantastic origin, from the earth sweating: 37.36). The whole is fascinating reading, and can be read in the Victorian translation of J Bostock and H T Riley, with notes reflecting the scholarship of the time, at www.perseus.tufts.edu/hopper/text?doc=plin.+nat.+toc. Scroll down at the left to book 37, and select chapters 11 and 12 (Bostock and Riley used a different section numbering).
3 Unless we assume very accurate information conveyed down the chain of merchants bringing it to the Adriatic and Mediterranean. At 37.43–5, he refers to the overland trade route trade from the Baltic through Austria to the Adriatic; see on F6.

Sotacos was a Greek mineralogist whose date is various stated as 4th–3rd century; it must have been after Pytheas made known the name for Britain, as discussed on F5. He is known almost entirely from Pliny. He was wrong in that Britain does not produce amber, but it is washed up on the east coast. The rest of §35 contains several difficulties which are discussed in Appendix 4 and the appended Note, with the following conclusions:

(1) Pliny's 'aestuarium' here means a 'tidal coast': Appendix 4 §§5 and 8, where its possible relation to Mela's tidal islands, F13, is noted.

(2) A tribe 'Guiones' is otherwise unknown, but it is the spelling in most Pliny MSS. One MS has 'Got(t)hones' (Goths), either an attempt to correct the text, or a copyist misreading an I for a T, an easy error in MS transmission. The name must go back to what Pytheas heard on the spot, and there are three possibilities: (a) he referred to the Inguaeones of F17, and wrote Ἰγγουϊώνεις, *Ingouioneis*, or similar, and it was miscopied (or misread) as Γουϊώνεις, *Gouïoneis*, by Pliny;[4] (b) Pytheas recorded them as Γουϊώνεις, *Gouïoneis*, perhaps because the locals did not stress the initial syllable in pronunciation, and Pliny preserves Pytheas' spelling; (c) he wrote Γοθώνεις, *Gothōneis*, and Pliny so found.[5] An analysis of the whole context shows that they are the Inguaeones: Appendix 4 §§8, 11 and Note to Appendix 4 §11.

(3) The phrase 'a Germanic tribe' does not help. As shown in that Note, the 'German' name was unknown in Pytheas' time, so it is not his phrase.[6] Pliny was variously explaining amber, and could have thought it desirable to locate this particular amber; if his source had 'Guiones', it would be a guess, given their proximity to the Teutoni. That comment equally applies if it was a later gloss that found its way into the text.

(4) The Metuonis coast is some part of the Baltic: Appendix 4 §§6–8, 11. It is a matter of comment that Pliny did not turn Pytheas' Greek stades into Roman miles; it is possible that he was troubled by the problem we face, of fitting it into the Codanus Bay. Note that the spelling is not certain: Pliny's MSS are broadly divided between 'Metuonidis' and 'Meconomon'. Both are transcriptions of Greek genitives, but the latter is plural, and place names in the plural are unusual. Either name, being in Greek form, must be from Pytheas; on this detail the MS tradition favours the singular form Metuonis, which most editors accept.[7]

---

4 Or when Pliny dictated the name to his shorthand writer, who misheard the first syllable.

5 In theory, Pliny's copy of Pytheas might have already had a misspelling, or was of poor legibility, or was mistranscribed by his secretary. The basic possibilities remain as stated. This point is not affected by the various spellings we have for the Goths, e.g. Γοῦται, *Goutai*, Ptol *Geog* 2.11.16, Γύθωνες, *Gythones*, *id* 3.5.8.

6 So far as is known, 'Germans' is not how they referred to themselves: Roseman (1994) 98.

7 See Detlefsen (1897) 191–3 (Guiones), 194 (Metuonis) for discussions of the problems. The best known plurals are Athens, Ἀθῆναι, *Athēnai*, and Patras, Πάτραι, *Patrai*.

(5) Resolving the conflict between the names 'Abalus' and 'Basilia' is in Appendix 4 §2; it is suggested the local name was 'Balchia' or 'Baltsia' or similar, which Pytheas rendered euphoniously as 'Basilia'.

(6) For the Teutoni, see the Note to Appendix 4, esp §11; they were basically in Jutland.

Pytheas was correct to the extent that amber is washed up by waves, some onto the island where he was (cf on F6). He was both right and wrong about its 'frozen sea' origin. He would assume that it was the same frozen sea which he had encountered in the far north; he could not know that the Baltic was a cul-de-sac and that it was melt-water from the Gulf of Bothnia, including that from rivers in Sweden and Finland, and the Gulf of Finland, and that this was why the islanders experienced stronger tides, with more amber washed up than at other times. He did not learn that the actual source of amber was the mainland some distance further east than where he was. If he hoped to initiate trade in amber, it seems that the locals disappointed him. He was misinformed or misunderstood about it being burnt. It does not burn at the temperatures of ordinary fires; it needs a fairly hot fire, about 300° C, to melt, at which point it decomposes. It is not suitable for fuel in the conventional sense. However, when put in a fire, it gives off an agreeable aroma. He may have found it being so used in houses, to sweeten the air; or perhaps he saw it being burnt in some ceremony.[8]

In a nutshell, this reference puts Pytheas in the Baltic, finding one source of amber, but disappointed if he hoped to break into its trade. See further Appendix 4 §7 for locating the island in the Baltic, and Appendix 6 §§14–16 for Pytheas' frozen sea.

F22 Schol in Ap Rhod 4.761–5a

'ὅθι τ' ἄκμονες Ἡφαίστοιο' αἱ τοῦ Αἰόλου νῆσοι ἑπτά. τούτων ἐν τῇ Λιπάρᾳ καλουμένῃ καὶ τῇ Στρογγύλῃ ὁ Ἥφαιστος δοκεῖ διατρίβειν, καὶ ἀκούεσθαι βρόμον πυρὸς καὶ ἦχον σφυρῶν. τὸ δὲ παλαιὸν ἐλέγετο τὸν βουλόμενον ἀργὸν σίδηρον ἐπιφέρειν καὶ τῇ ἑξῆς ἡμέραι ἢ ξίφος λαμβάνειν ἢ εἴ τι ἄλλο ἤθελεν κατασκευάσαι, καὶ τὸν μισθὸν ἀπεδίδου. ταῦτα ἱστορεῖ Πυθέας ἐν Περιόδῳ γῆς, καὶ τὴν θάλασσαν λέγων ζεῖν. εἰσὶν δὲ αἱ τοῦ Αἰόλου νῆσοι ζ αἵδε· Στρογγύλη Εὐώνυμος Λιπάρα Ἱερά Διδύμη Ἐρικώδης Φοινικώδης.

'Where the anvils of Hephaistos': there are seven islands of Aiolos. Hephaistos appears to spend his time on the ones called Lipara and Strongyle; and the noise of fire is heard, and the echo of hammers. In olden times it was said that one could bring iron ore, and the next day collect a sword or anything else that had been made, and pay for it. Pytheas says this in his World Guide, *adding that the sea boils. These are the 7 islands: Strongyle, Euonymos, Lipara, Hiera, Didume,*

---

8 A page or so after his citation of Philemon at *NH* 37.33 (see n 2), Pliny adds that Philemon said that white amber does not burn, 37.36.

Ericodes, Phoenicodes [Stromboli, Panarea, Lipari, Vulcano, Salina, Alicude, Philicudi].

Apollonios of Rhodes (third century) wrote a long epic poem about the voyage of the *Argo* which enjoyed great popularity and became the standard version of the myth. It soon attracted *Scholia*, commentaries, written in Alexandria between the late second century (BC) and the first century AD;[1] a copy of Pytheas' book would be in the library there. At this point in the poem, the goddess Hera tells the messenger Iris to go where Hephaistos' anvils are, and tell him to keep quiet until the *Argo* has passed; and then to Aeolos, and tell him to give the *Argo* a fair west wind.[2] Our text is a commentary on the mention of anvils. In the Greek pantheon, Hephaistos was the god of fire and metalworking, typically associated with an island, usually the Aeolian islands off Sicily; his later identification with the Roman god Vulcan also put him under Mt Etna on the Sicilian mainland. The translation includes the modern names.[3]

The Pytheas reference is puzzling; a resolution ties up with the reference to Pytheas' book in F30 §6. In a context of hostile exaggeration, it is there asserted that he sailed 'from Cadiz to the Don', but along the coast of the Ocean, not the Mediterranean. Even so, coupled with the title offered here, Periodos Gēs ('World Guide'), different to the 'On the Ocean' of F3 and F7, it raises the question whether Pytheas wrote a second book about the Mediterranean and Black Sea. But all our references to Pytheas imply just one book, about the unknown Ocean and the north. Also, it is difficult to see why the man who had explored distant and unknown areas should want to write a guide to well-known waters, as supplement to his Ocean if not a second book. See further on F30 §6 and the Coda, p 177. But if he was never on the Aeolian islands, how do we explain this reference? By the time the commentary on Apollonios was written, World Guide had become a generic name for travel books, and its use here could be carelessness by the scholiast. However, there is another difficulty. A myth about leaving old iron to be reworked into something new is recorded elsewhere in slightly different terms. It is just the kind of story commonly attached to a place in travel guides, a myth associated with a place: see on F9. Equally it is an odd story for Pytheas to have included; all our other references suggest that he recorded scientific observations, or facts such as headlands, not myths. It is, therefore, feasible that the scholiast confused Pytheas with a writer of an actual World Guide, and F22 is not from Pytheas. In a world where it was very difficult to look up a particular passage in a papyros roll, citations from other authors were usually from memory; errors of memory were easy to make.

---

1 See paragraph F in the *BNP* sv Apollonius Rhodius. There are four scholia on this passage, conventionally lettered a to d in modern editions; our excerpt is a.
2 In book 10 of the Odyssey, Aeolos is the keeper of the winds, living on the island Aeolia; the Aeolian islands name may reflect that.
3 There are now eight islands, as the islet of Basiluzzo between Stromboli and Panarea is included.

If it was from him, he could perhaps have mentioned the Aeolian islands and their volcanic nature ('the sea boils') as the opposite of the frozen sea; the sea boils when Hephaistos is busy, as in the story about him reworking old iron. Other explanations are unconvincing. Pytheas cannot have been comparing the islands to Iceland; even if we accept that Thule was Iceland (Appendix 6 §5), it would be an astonishing coincidence if those who had visited it had been there when there was not only an eruption, but one close enough to the coast to cast molten lava into the sea and cause it to 'boil'. More plausible, but still unlikely, is the suggestion that when in Britain he encountered hot springs, such as at Bath, and mentioned the boiling sea in that connection: Roseman (1994) 115.

F23 Steph Byz sv Ὠστίωνες ≈ Aelius Herodianus, *De Pros Cath* 3.1.19

Ὠστίωνες ἔθνος παρὰ τῷ δυτικῷ Ὠκεανῷ, οὓς Κοσσίνους Ἀρτεμίδωρός φησι, Πυθέας δ᾽ Ὠστιαίους· τούτων δ᾽ ἐξ εὐωνύμων οἱ Κόσσινοι λεγόμενοι Ὠστίωνες, οὓς Πυθέας Ὠστιαίους προσαγορεύει.

Ōstiones: Ōstiōnes, a people by the western Ocean, whom Artemidoros calls Cossinoi, Pytheas Ōstiaioi: to the left of these are the Cossini [also] called Ōstiōnes, whom Pytheas calls Ōstiaioi.

Herodianus was a grammarian of the second century AD; he noted these names in a passage listing the correct spelling of the genitive plural of ethnic names. Clearer from our point of view is the entry in the list of ethnic names, *Ethnica*, of Stephanus of Byzantium, c 6th century AD;[4] but Herodianus shows that these forms of the name already existed in his time, some 150 years after Strabo's death. Pytheas met this tribe, but his spelling is difficult to recover because Strabo's MSS offer variant readings (F25, F26, F37). The text shows how foreign names could circulate in a variety of forms; *Ōstiōnes* is an 'odd' form, with its not very common ending *-ōnes*. The problem is discussed in Appendix 5, with the conclusion that *Ōstiaioi* is close to how Pytheas spelled it, but it has lost its middle consonant; may have written *Ōstidaioi*.

For Artemidoros, see p xix. His 'Cossini' are otherwise unknown; it suggests that he was writing before knowing of Caesar's *Bellum Gallicum*; if he had accessed it, we would expect a spelling closer to Caesar's 'Osismi'. Either the name derives from Pytheas, mangled from mishearing or miscopying, or he picked up some mariner's or trader's story, perhaps in his native Ephesos, in whose mouth the name was no longer close to the local pronunciation.

---

4 Herodianus says that the genitive plural of *Ostiones* is *Ostiōn* and not the technically correct *Ostiōnōn*. Note that Stephanus' *Ethnica* survives as a later epitomy, but the entry is adequate for our purposes.

Introductory note to the Strabo references

All the remaining references come from Strabo. Most of them make pejorative references to Pytheas, and it is essential to understand Strabo's standpoint. He had a dogmatic adherence to the Stoic view of the world: the earth was divided horizontally into five zones, first recorded in surviving literature by Aristotle (p 20). The equatorial zone was uninhabitable because of the extreme heat; the two zones near the poles were uninhabitable because of the extreme cold. Only the two middle zones were temperate and habitable, the *oikoumenē* ('lived in'), i.e. habitable zone, which included Greece, and an assumed similar zone south of the equator. Strabo believed that Ireland was at the northern limit of the *oikoumenē* (F25, F31); therefore Pytheas, who described life in the far north, had to be lying.

Books 1 and 2 of his *Geography* are mainly a review of his predecessors. He starts with Homer (1.2), and among others he looks at Eratosthenes, hence F24–F26; Hipparchos, so F27–F28; Posidonios, p xx, and Polybios, F30. His approach is: where you say 'X' and I, Strabo, say 'Y', it is you who are wrong. He concludes book 2 with his own geography of the world, 2.5, from which F31 to F33 come.[1] He does not treat Pytheas as a predecessor, and it is probable that his references to Pytheas are from the intermediate author he is citing (but see on F38). Books 3 to 17 cover individual areas of the known world in such detail as he could muster. His distances do not always tally.[2] He struggled with mathematics; he had a practical approach to geography, and was not troubled if he came up with slightly different figures for the same distance; when he could not follow someone's calculations, he retreats into 'ill-directed sarcasm'.[3] F24–F26 are a continuous text in Strabo; they are split here for convenience.

F24 Strabo 1.4.2

Ἑξῆς δὲ τὸ πλάτος τῆς οἰκουμένης ἀφορίζων φησὶν ἀπὸ μὲν Μερόης ἐπὶ τοῦ δι' αὐτῆς μεσημβρινοῦ μέχρι Ἀλεξανδρείας εἶναι μυρίους, ἐνθένδε εἰς τὸν Ἑλλήσποντον περὶ ὀκτακισχιλίους ἑκατόν, εἶτ' εἰς Βορυσθένη πεντακισχιλίους, εἶτ' ἐπὶ τὸν κύκλον τὸν διὰ Θούλης (ἥν φησι Πυθέας ἀπὸ μὲν τῆς Βρεττανικῆς ἓξ ἡμερῶν πλοῦν ἀπέχειν πρὸς ἄρκτον, ἐγγὺς δ' εἶναι τῆς πεπηγυίας θαλάττης)

---

1 Books 1 and 2 are usefully summarised in the Wikipedia entry 'Geographica'.
2 It is probable that he died before finalising his work, and he might otherwise possibly have noted conflicting distances and attempted to resolve them. But it is also clear that his sources often offered different distances between the same two places.
3 'He begins to breathe uneasily as soon as he enters the rarified atmosphere of mathematical geography': Dicks (1960) 162, with detailed instances of Strabo's inability to cope; *ibid* 171. His practical approach, *ibid* 191; he says he is writing for statesmen who do not want troubling with astronomy: 2.5.34.

ἄλλους ὡς μυρίους χιλίους πεντακοσίους. ἐὰν οὖν ἔτι προσθῶμεν ὑπὲρ τὴν Μερόην ἄλλους τρισχιλίους τετρακοσίους, ἵνα τὴν τῶν Αἰγυπτίων νῆσον ἔχωμεν καὶ τὴν Κινναμωμοφόρον καὶ τὴν Ταπροβάνην, ἔσεσθαι σταδίους τρισμυρίους ὀκτακισχιλίους.

Next, to establish the breadth of the oikoumenē *[inhabitable world], he [Eratosthenes] says that, from Meroë along this longitude to Alexandria is 10,000 (stades) [1,135 miles, 1,815 km]; from there to the Hellespont about 8,100 [920 miles, 1,470 km]; then to the Borysthenes (the mouth of the Dnieper) 5,000 [570 miles, 910 km]; then to the latitude through Thule, which Pytheas says is six days' sail from Britain to the north, and is near the frozen sea, about 11,500 more [1,310 miles, 2,095 km]. So if we add another 3,400 stadia [385 miles, 615 km] beyond Meroë [i.e. to the south], in order to include the island of the Egyptians, the Cinnamon-Producing Land, and Taprobane, there will be 38,000 stades [4,320 miles, 6,910 km].*

This is a simple text, but confusing to read without knowing either its context, or where the various places are. Strabo has just said that Eratosthenes' dimensions for the *oikoumenē*, the inhabited world, are not entirely correct. This text and F25 deal with the width; F26 with the length. Alexander's conquest of the east had produced many distances for that part of the world, of varying degrees of accuracy,[1] which, coupled with earlier information, enabled Eratosthenes either to produce or at least describe a map which included the longitude in this text, and several latitudes; one mentioned here, and others in F25, F27, F28 and F31. The longitude became standard in ancient geography, though usually expressed from Alexandria as via Rhodes and Byzantium (e.g. F32). It is convenient to clarify it in a slightly different order to Strabo.

Meroë, now an archaeological site in the Sudan, was said to be the capital of the Ethiopians (Strabo 17.3.10).[2] The southern limit of the *oikoumenē* was the latitude through three places further south. The 'island of the Egyptians' is usually identified with the place mentioned by Herodotos at 2.30.1, the area between the White and Blue Niles a little to the south of Khartoum (Asheri *et al* (2007) 260; in more detail Lloyd (1994) 126–8).[3] Meroë is some 150 miles (240 km) north-east of Khartoum by road, which follows the Nile, and about 120 miles (190 km) as the crow flies. Both are rather less than the 3,400

---

1 Even in Eratosthenes' time it was known that information about India from men associated with Alexander's conquests was of variable quality: Strabo 2.1.4–9.
2 The archaeological site of Meroë is at 16° 56′ N, 33° 44′ E, 29 miles, 46 km, north-east of Shendi, variously said to be at the villages of Kabashiya and Al-Bagawiya.
3 The area immediately south of the confluence of the White and Blue Niles is or was called 'Gezirah', i.e. 'island', though the modern Gezira or Al-Jezirah, one of the states of modern Sudan, is somewhat further away. There is no other obvious island or locality that suits.

stades here, 3,000 stades (385, 340 miles; 615, 545 km) elsewhere; Herodotos' distance is also problematic.[4] The Cinnamon country is the coast of Ethiopia, now Somalia. In fact spices came from the east, were transhipped in Somalia, and reached the Mediterranean up the Red Sea;[5] though we cannot assume that Greeks in Eratosthenes' time knew the actual shape of the Horn of Africa.[6] Taprobane is Sri Lanka.[7] Eratosthenes erred in putting a latitude through these three places. In round figures, the Somali coast is about 350 miles, 560 km, south of the latitude through Khartoum; Sri Lanka is further south still, the northern tip 130 miles, 210 km, and Colombo 350 miles, 560 km.[8] Strabo repeats this latitude with slightly different figures at 2.1.13, 14, 17; 2.2.2; 2.5.7, 8, 35, which confirm that it was Eratosthenes'; he addresses its difficulties in F27.[9]

Moving north, the next place here is Alexandria. This stretch, here 10,000 stades, is usually expressed as 5,000 stades from Meroë to Syene, Aswan, and 5,000 stades from there to Alexandria (Strabo 2.2.2, 2.5.7, just before F31). The south to north alignment of the Nile must have long been assumed, but it had recently been explored under the Ptolemies to Meroë. The evidence suggests that the distances recorded were based not on measurement or formal survey, but on how many days' journey were needed for each stage, by land or water, and translated into stades by some formulae, and rounded up or down.[10] From Syene to Alexandria,

---

4 3,000 stades at Strabo 2.1.14, 2.5.7, 2.5.8, 2.5.35. Herodotos described it as 56 days' journey from Meroë, the difficulties of which are discussed in Asheri *et al* and, in more detail, Lloyd *ad loc*.

5 It was also transhipped at an Arabian port, e.g. Aden, whence it came overland up Arabia; hence Arabia was thought to be its source, e.g. Herodotos, p 8 n 7; Pliny *NH* 10.97, 12.82, 87–93; at *NH* 12.40–42 he has stories about cinnamon as fantastic as those about amber noted on F21. Strabo 16.4.14 says that cinnamon comes from Somalia but more abundantly from the interior; generally *BNP* sv Kinnamomophoros Chora.

6 There would be some knowledge in Alexandria from traders and maritime exploration under the Ptolemies, but our first surviving written record is the *Periplus of the Red Sea*, mid first century AD, which referred to the Elephant Promontory, §11, Ras Filuch or El-Fil per Barrington 4; the two trading centres (emporia) of Mosylon and Aromatica (Spices), §§10, 12 imply places known for some time. Aromatica cannot be securely located; Mosylon is put at Bandar Kassim (Bosaso) or Qassim.

7 There was some uncertainty about Sri Lanka; it is 'very far south' of India in Strabo 2.5.14, and seven days' sail at 15.1.14.

8 Khartoum is at 15° 30' N; Qandala, Somalia, is south-east at 11° 28' N (Aden is at 12° 48'), while Colombo, Sri Lanka, more so, 6° 56' N.

9 Strabo also notes the three places individually when he describes Asia and Africa: Egyptian Island, 17.1.2; Cinnamon country, 16.4.4, 14, 20; 17.1, 5; Taprobane 2.1.14; 15.1.14, 15.

10 The main passages are Pliny *NH* 5.59, 6.183, 191, 194; Strabo 2.1.20, and giving us the names of Dalion, Aristocreon, Bion, Simonides the younger, Philon, and Basilis (*FGrH* 666–670, 718); perhaps we should add Simmias, Diodoros 3.18.4. Pliny's numbers have suffered some distortion in transmission, and his citations vary between x [Roman] miles and y days' journey. Diller (2010) 157 suggests that Eratosthenes accessed the *Aithiopica* of Philon (*FGrH* 670 F2 = Strabo 2.1.20), covering the Sudan rather than modern Ethiopia. Pliny *NH* 6.183 notes that Timosthenes, noted on F9, reported 60 days' sailing from Syene to Meroë, similar to the 56 days Herodotos offered at Hdt 2.29.

the Nile does flow in a reasonably straight line and roughly south to north; 5,000 stades as a round figure (570 miles, 910 km) is an acceptable distance: by the modern road, which mainly runs close to the Nile, it is 680 miles, 1,090 km. But between Meroë and Syene, it forms a large loop. The Ptolemaic explorers seem to have noted this; Eratosthenes knew of it, as is clear from Strabo 17.1.1–2. By the modern road, also more or less close to the river, it is 830 miles, 1,330 km. But Eratosthenes was here defining a longitude, and he could calculate or at least assess the straight line distance between the two places. On that basis, his 5,000 stades is also not inaccurate as a round figure: the two are about 500 miles, 800 km, distant as the crow flies. As noted above, a longitude with 5,000 stades from Meroë to Syene and 5,000 stades from Syene to Alexandria became generally accepted.

Because the sun at Syene at the solstice was directly overhead (pp xix–xx), Eratosthenes could place it on what he called the summer tropic, which for him meant 24° north of the equator: 4/15 of 90°, and so 16,800 stades north of the equator.[11] It followed that if Meroë is 5,000 stades south of Syene, and the Cinnamon country is 3,400 or 3,000 stades south of Meroë (see above), then the Cinnamon country is 8,400 stades or 8,800 stades from the equator (955, 1,000 miles; 1,530, 1,600 km). This 16,800 stades is noted also on F25 and F32.

From Alexandria northwards, Eratosthenes offered 8,100 stades to the Hellespont, and a further 5,000 stades to the mouth of the Dnieper. The Greek just says 'Borysthenes', the Dnieper; Strabo regularly uses the name when he means its mouth (so also F25, F27).[12] Later geographers expressed this as via Rhodes, a major shipping hub, and Byzantium: see on F32 for the various distances that are given. It is broadly south to north; there various routes to the Hellespont from Rhodes, along the west coast of Turkey or through the offshore islands. Both for that stretch, and then from the Hellespont to Byzantium and along the coasts of Bulgaria and Romania to the Dnieper (and the nearby Greek colony of Olbia), while a sea-captain would know from the sun that individual stretches took him now north-east, now north-west, as he negotiated

11 Strabo 2.5.7, just before F27; Strabo is here using Hipparchos, who himself basically adopted Eratosthenes' figures. We must not confuse ancient tropics with ours. Our Tropic of Cancer is currently at about 23° 26' N; Syene, Aswan, is at 24° 05' N; Capricorn is in the southern hemisphere. We are concerned with the tropics of Greek astronomers in the northern hemisphere, at 24° from equator and pole; both can be described as summer tropics.
12 It is odd that the river mouth was used, rather than the nearby port and town, the Greek colony of Olbia (modern Parutyne, Ukraine), about 15 miles, 24 km, north of the mouth of the Dnieper on the west bank of the Hypanis, the Bug. Roller (2010) 10 suggests that Borysthenes was (also) the collective name for the cluster of Greek settlements in the area. Herodotos, describing Scythia, names the river, and the locals (Borysthenites), but also calls the town 'Borysthenes': 4.24, 78.5; perhaps it was originally so known.

the coasts and islands and allowed for the wind, his overall progress was from south to north. As round figures, Eratosthenes' 8,100 stades and 5,000 stades are reasonable.[13]

From the Dnieper to Thule, Eratosthenes (although himself a Stoic) accepted Pytheas' Thule. His distance of 11,500 stades, is surprisingly accurate.[14] Strabo could not accept Thule, and attempts to discredit Eratosthenes in F25. The frozen sea, here 'near' Thule, is more precisely located by Pliny, F19, as one day's sail distant. It is further discussed in the light of other references to it: see Appendix 6 §§14–16.

F25 Strabo 1.4.3–4

[3] Τὰ μὲν οὖν ἄλλα διαστήματα δεδόσθω αὐτῷ· ὡμολόγηται γὰρ ἱκανῶς· τὸ δ' ἀπὸ τοῦ Βορυσθένους ἐπὶ τὸν διὰ Θούλης κύκλον τίς ἂν δοίη νοῦν ἔχων; ὅ τε γὰρ ἱστορῶν τὴν Θούλην Πυθέας ἀνὴρ ψευδίστατος ἐξήτασται, καὶ οἱ τὴν Βρεττανικὴν καὶ Ἰέρνην ἰδόντες οὐδὲν περὶ τῆς Θούλης λέγουσιν, ἄλλας νήσους λέγοντες μικρὰς περὶ τὴν Βρεττανικήν· αὐτή τε ἡ Βρεττανικὴ τὸ μῆκος ἴσως πώς ἐστι τῇ Κελτικῇ παρεκτεταμένη τῶν πεντακισχιλίων σταδίων οὐ μείζων καὶ τοῖς ἄκροις τοῖς ἀντικειμένοις ἀφοριζομένη. ἀντίκειται γὰρ ἀλλήλοις τά τε ἑῷα ἄκρα τοῖς ἑῴοις καὶ τὰ ἑσπέρια τοῖς ἑσπερίοις, καὶ τά γε ἑῷα ἐγγὺς ἀλλήλων ἐστὶ μέχρις ἐπόψεως, τό τε Κάντιον καὶ αἱ τοῦ Ῥήνου ἐκβολαί. ὁ δὲ πλειόνων ἢ δισμυρίων τὸ μῆκος ἀποφαίνει τῆς νήσου, καὶ τὸ Κάντιον ἡμερῶν τινων πλοῦν ἀπέχειν τῆς Κελτικῆς φησι· καὶ τὰ περὶ τοὺς Ὠστιδέους δὲ καὶ τὰ πέραν τοῦ Ῥήνου τὰ μέχρι Σκυθῶν πάντα κατέψευσται τῶν τόπων. ὅστις οὖν περὶ τῶν γνωριζομένων τόπων τοσαῦτα ἔψευσται, σχολῇ γ' ἂν περὶ τῶν ἀγνοουμένων παρὰ πᾶσιν ἀληθεύειν δύναιτο. [4] Τὸν δὲ διὰ τοῦ Βορυσθένους παράλληλον τὸν αὐτὸν εἶναι τῷ διὰ τῆς Βρεττανικῆς εἰκάζουσιν Ἵππαρχός τε καὶ ἄλλοι ἐκ τοῦ τὸν αὐτὸν εἶναι καὶ τὸν διὰ Βυζαντίου τῷ διὰ Μασσαλίας· ὃν γὰρ λόγον εἴρηκε τοῦ ἐν Μασσαλίᾳ γνώμονος πρὸς τὴν σκιάν, τὸν αὐτὸν καὶ Ἵππαρχος κατὰ τὸν ὁμώνυμον καιρὸν εὑρεῖν ἐν τῷ Βυζαντίῳ φησίν. ἐκ Μασσαλίας δὲ εἰς μέσην τὴν Βρεττανικὴν οὐ πλέον τῶν πεντακισχιλίων ἐστὶ σταδίων. ἀλλὰ μὴν ἐκ μέσης τῆς Βρεττανικῆς οὐ πλέον τῶν τετρακισχιλίων προελθὼν εὕροις ἂν οἰκήσιμον ἄλλως πως (τοῦτο δ' ἂν εἴη τὸ περὶ τὴν Ἰέρνην), ὥστε τὰ ἐπέκεινα, εἰς ἃ ἐκτοπίζει τὴν Θούλην, οὐκέτ'

---

13 Modern distances are: Alexandria to Istanbul, 832 miles, 1,331 km, direct, 380 + 486 = 866 miles, 608 + 778 = 1,386 km, via Rhodes; Istanbul to Odessa (the nearest modern port) 394 miles, 630 km. Allowing for the limitations of ancient navigation and the need to keep close to the coast, Eratosthenes' equivalent of 920 + 570 miles, 1,472 + 412 km, is realistic.
14 Any measurement on a modern map for such a distance will depend on the distortion arising from the method used to represent the curved surface of the earth on a flat piece of paper (or computer screen); but the 1,305 miles, 2,090 km, equivalent of 11,500 stades is close to the mark. See on F25 for Eratosthenes' calculation.

οἰκήσιμα. τίνι δ' ἂν καὶ στοχασμῷ λέγοι τὸ ἀπὸ τοῦ διὰ Θούλης ἕως τοῦ διὰ Βορυσθένους μυρίων καὶ χιλίων πεντακοσίων, οὐχ ὁρῶ.

[3] *Let his [Eratosthenes'] other distances be accepted, for they are generally agreed; but who could sensibly believe the distance from the mouth of the Dnieper to the latitude through Thule? For Pytheas, the man who reports on Thule, has been proved to be an arch-liar. Those who have seen Britain and Ireland say nothing about Thule, though they mention other small islands around Britain. Britain itself more or less lies alongside Keltikē for a distance of no more than 5,000 stades [570 miles, 910 km], with their ends opposite each other. For its eastern ends lie opposite the eastern ends of Keltikē, and similarly with their western ends; and the eastern ends at any rate are close enough to be mutually visible, Kantion and the mouths of the Rhine. But he [Pytheas] declares that the length of Britain is more than 20,000 stadia [2,270 miles, 3,630 km], and that Kantion is several days' sail from Keltikē; and as to both the Ostideoi and what is beyond the Rhine as far as Scythia he always tells lies about the places. So a man who has lied to such an extent about known regions would hardly be able to speak truthfully about places that are unknown to everyone. [4] Hipparchos and others infer that the latitude of the mouth of the Dnieper is the same as that of Britain, from the fact that the latitude of Byzantium compares to that of Marseille; for the ratio of the gnomon to its shadow at Marseille which he [Pytheas] stated is the same as what Hipparchos says that he found at Byzantium at the same time of year. From Marseille to the middle of Britain is no more than 5,000 stades [570 miles, 910 km]. But in fact going on from the middle of Britain for no more than 4,000 stades [4557 miles, 730 km] you would find a region that is only just habitable: this would be in the region of Ireland. Therefore what is further on, where he [Eratosthenes] locates Thule, are no longer habitable. By what guessing he could say that the distance from the latitude through Thule to that through the Dnieper is 11,500 stadia [1,305 miles, 2,090 km], I do not see.*

This text immediately follows his description of Eratosthenes' latitude, F24. He now mounts a two-pronged attack on him for extending it northwards to the latitude through Thule; Pytheas lied about the north, so Eratosthenes was wrong to believe him (§3); and the distances prove this (§4).

§3 Strabo's attack here is plausible rhetoric and not sound reasoning; cf Radt's phrase, noted on F29, of how he loses himself in poor arguments. He starts on a high note; his 'arch-liar', ψευδίστατος, *pseudistatos*, is the superlative of ψευδής, *pseudēs*, 'false', 'lying'. The 'proof' is that 'those who have seen Britain and Ireland' do not mention Thule. But that is special pleading. We know of no one in Strabo's time who visited Britain after Pytheas except Caesar, who was only in the south-east, and nowhere near Scotland (he had heard of Ireland, BG 5.13.2). If Timaios noted Thule, Diodoros (e.g. F4–F5) ignored it. Anything that circulated orally about Britain would be from traders; trading contacts were with ports on the south coast. Caesar, *BG* 4.20, said that merchants in Gaul could only tell him about the very south of England. Only a visitor to the Northern Isles, or perhaps the north of Scotland, would be likely to hear about Thule; in the south, they would be unlikely to have heard even about the Northern Isles, much less Thule. Indeed, for his geography of Britain and Ireland, Strabo could only offer a modest description (4.5.1–5: see on F38).

Strabo is ambiguous as to the existence of Thule: is it an imaginary island, or is there an uninhabited island in the far north which Pytheas represented as inhabited? Pytheas either said that it was inhabited, or implied as much: Eratosthenes so assumed when calculating its latitude; for him it was just in the *oikoumenē*, the habitable zone; cf Appendix 6 §12. Here, having put Ireland at the limit of the *oikoumenē*, he says that τὰ ἐπέκεινα, *ta epekeina,*'what is further' or 'beyond', is uninhabitable (§4). This could mean either the Ocean, or an uninhabited island in it. In F31 he notes that Pytheas said that Thule is where the summer tropic is the same as the arctic circle. He has some respect for Pytheas' astronomy (F38, F39) that perhaps explains his wording. In F38, he says in terms that others mention Thule, but it is a fabrication of Pytheas. As elsewhere, Strabo's advocacy is always selective.[1]

Like Mela, F12, and Pliny, F18, Strabo puts Kent opposite the Rhine (also at 4.5.1, noted on F27). Pytheas did not say this, and Caesar 5.13.1 placed it correctly opposite Gaul. The Rhine was not known until the second century (introduction and §1 to the Note to Appendix 4). But once known, it became the eastern edge of Keltikē. Since Britain extended opposite Keltikē, if Keltikē extended to the Rhine, the eastern tip of Britain, Kantion, was also opposite the Rhine; see also Appendix 7. Strabo's 5,000 stades for the south coast is more realistic than Pytheas' 7,500 stades in F4, though, as there noted, he also had lower figures of 4,300 and 4,400 stades (490 and 500 miles, 785, 800 km); further, he treats this as its longest side, as to which see Appendix 7. He now misrepresents Pytheas; as we see in F4, Pytheas put the *west* coast at 20,000 stades. Strabo would have been on surer ground if he had attacked Pytheas' 7,500 stades for the south coast. He also probably misrepresents him with his Kent being several days from Keltikē, and not just because of the implication that since the two coasts were mutually visible, anyone who said otherwise was lying. His 'several' is the indefinite τις, *tis* (here in the genitive plural); all other distances attributed to Pytheas are definite numbers, e.g. the four days to Cornwall from (probably) the Loire estuary (see on F4). In any event, Pytheas sailed along the English Channel, and would know that at one point they are mutually visible.[2] Either Strabo misapplied the France to Cornwall figure here, or Pytheas recorded some other distance three or four days from France to Kent.

---

1 Virgil could only have spoken of Ultima Thule if it was widely known from earlier writings: p 1.
2 Diodoros accessed the detail that Kent was 100 stades (11 miles, 18 km) from the Calais area, most probably from Pytheas via Timaeios (see on F4); Strabo did not know this or overlooked it; it is at best uncertain that Strabo ever read Pytheas' actual book.

The next sentence is particularly problematic: (a) who are the Ostideoi whom Strabo suddenly mentions, and (b) why does he add the Scythia reference? His readers might be puzzled as to whom these Ostideoi are. Common sense would suggest that they are the similarly named tribe in Brittany he is about to mention in F26; he had them in mind here also, and keen to pile on his rhetoric forgot that he has not already mentioned them. Unfortunately, in F26 their name appears with different spellings in his MSS, and this had led Lasserre (1963) to propose that these Ostideoi are a different tribe, on or just across the Rhine. But we must note how Strabo expresses himself. He joins his 'Ostideoi' to what is beyond the Rhine not just with καί . . . καί, *kai . . . kai*, the usual Greek for 'both . . . and', but with the strong joining particle δὲ καί . . . καί, *de kai . . . kai*, with the connotation 'not just the Ostideoi, but also everything up to Scythia'. His language shows that he associates the Ostideoi with Keltikē, not with the Rhine or Scythia.[3] He throws in Scythia as an additional area for which Pytheas told lies, and helps to place the latter in the Baltic (Appendix 4 §§7–10). Further, there is no trace of this Germanic Ostideoi tribe in any other author where Germanic tribes are noted, and the suggestion that they could be identified with the Inguaeones (Lasserre (1963) 113 n 18) is clutching at straws. For further analysis showing that these Ostideoi are the Brittany tribe, see Appendix 5.

He concludes with a rhetorical but quite inaccurate flourish: in view of the known facts about which Pytheas lied, therefore he lied about areas 'unknown to everyone': everything up to Scythia, and for Strabo, Scythia began somewhere beyond the Elbe: see Note to Appendix 4, §3. Pytheas sailed from the Calais area to Jutland and into the Baltic (Coda pp 175–6): all lies, according to Strabo, who unfortunately from our perspective failed to say what those lies were. Further, on any view, he was being economical with the truth. It is clear from Mela, probably born in Strabo's lifetime, in F13, and Pliny, F17, that stories about the Baltic had long circulated; and Strabo glosses over his own rather limited knowledge of Germany (see the earlier Note).

§4 Strabo refers to the Marseille-Byzantium latitude elsewhere: F27, F31, F32, and 2.4.3. As in F31, he notes that the two *gnomon* readings were not quite the same; here he says that astronomers εἰκάζουσι, *eikazousi*, 'compare' the two latitudes. It was probably a spring equinox reading: Appendix 2 §2. By modern standards, Hipparchos erred, and this affects any attempt to correlate other distances and calculations which depend on it with modern figures.[4] Greek astronomers

---

3 It has been proposed that Pytheas treated this Ost- tribe as living as far east as Frisia or Jutland (Lasserre (1963) 111 n 12; Murphy (1977) 53); but this depends on treating Avienus' *Ora maritima* 129–45 as based on the alleged *Massiliot Periplus*, as to which see Appendix 1.

4 See, e.g., Roseman (1994) 43, for modern figures. For *gnomon* readings see Appendix 2 §3. Marseille is at 43° 18′; Byzantium, Istanbul, is at 41° 01′. Note that Radt *ad loc*, vol 5 (2006) 174 argues that the 'he' in 'he stated is the same' is Eratosthenes, not Pytheas, as the passage as a whole is an attack on Eratosthenes. But the context, as elsewhere, is that Hipparchos used Pytheas' *gnomon* reading, not Eratosthenes'. Roller (2018) 64 is badly phrased, as it appears to say that it was Pytheas who made the comparison with Byzantium.

assumed that small differences would not matter (p 21 n 12: perhaps accepting that they could not work to finer limits). So here; his summer daylight figure of 15¼ hours for this latitude (see on F28) may be compared to the actual figures of 15 hrs 26 min at Marseille and 15 hrs 8 min at Istanbul.

His argument is formally correct but illogical. He is dogmatic that Ireland is the limit of the *oikoumenē*, the habitable zone: he begins the sentence with ἀλλὰ μὴν, alla mēn, 'indeed' with the connotation 'it is definitely so'.[5] If that limit is 9,000 stades, 1,025 miles, 1,640 km, north of Marseille, then anywhere 11,500 stades, 1,305 miles, 2090 km, north of Marseille must be uninhabitable. For Strabo, if Thule existed, it was uninhabitable (see above), and logically he should have said that Eratosthenes was wrong to extend the *oikoumenē* so far north. Instead he claims not to understand the calculation; it reveals his weakness in mathematics: the calculation is simple. Eratosthenes had established the circumference of the earth at 252,000 stades (pp xix–xx), so the quarter from the equator to the north pole was 63,000 stades. He would see that Pytheas' summer tropic/arctic circle detail for Thule, coupled with the long daylight there (Appendix 2 §8) meant that Thule in the north was like Syene in the south; it was at the same distance from the north pole as Syene was from the equator, and so on the arctic circle in our sense. The latter distance is 16,800 stades, as noted on F24. Given the other distances on the longitude he had just detailed in F24, it followed that 11,500 stades was, in round figures, the distance from the mouth of the Dnieper to Thule.[6] With his poor mathematics, dogmatic views about the cold north, and disbelief in Thule anyway, Strabo could not or would not understand Eratosthenes' simple calculation; hence his 'I do not see'.[7] Ironically, he offers the Syene to equator distance again in F32.

There are two further points arising, discussed in Appendix 7. We get mixed messages as to whether he accepts the accuracy of the Marseille-Byzantium latitude. Here, in F27, and at 2.4.3, he accepts it; he uses it here to attack the Thule latitude; he attempts to discredit it in F31. Secondly, he thought that Ireland lay north of Britain, not west; that makes it impracticable to compare modern distances with those here. As the crow flies, London is about 620 miles, 990 km, from Marseille, not inconsistent with 5,000 stades, 570 miles, 910 km; but that may just be a coincidence. But Strabo's view of north-west Europe was incorrect: see Appendix 7 and maps 3a and 3b, p 161.

---

5 Denniston (1954) 341–7.

6 From F24: equator to Syene, 16,800 (stades); Syene to Alexandria, 5,000; Alexandria to the mouth of the Dnieper, 13,100. A further 11,500 to Thule and 16,800 from there to the north pole actually totals 63,200, but there were variations in the Alexandria and Rhodes to the Dnieper distances (see p xvii); cf on F27, and F31.

7 Pliny, *NH* 2.246, reported that Isidoros, for whom see on F18, put Thule 1,250 (Roman) miles from the Black Sea, specifically from the mouth of the Don. He called it a guess; at 8 stades to the mile, for which see p xvi, Isidoros would have written 10,000 stades.

F26 Strabo 1.4.5

Διαμαρτὼν δὲ τοῦ πλάτους ἠνάγκασται καὶ τοῦ μήκους ἀστοχεῖν. ὅτι μὲν γὰρ πλέον ἢ διπλάσιον τὸ γνώριμον μῆκός ἐστι τοῦ γνωρίμου πλάτους, ὁμολογοῦσι καὶ οἱ ὕστερον καὶ τῶν παλαιῶν οἱ χαριέστατοι· λέγω δὲ τὸ ἀπὸ τῶν ἄκρων τῆς Ἰνδικῆς ἐπὶ τὰ ἄκρα τῆς Ἰβηρίας, τοῦ ἀπ᾽ Αἰθιόπων ἕως τοῦ κατὰ Ἰέρνην κύκλου. ὁρίσας δὲ τὸ λεχθὲν πλάτος, τὸ ἀπὸ τῶν ἐσχάτων Αἰθιόπων μέχρι τοῦ διὰ Θούλης ἐκτείνει πλέον ἢ δεῖ τὸ μῆκος, ἵνα ποιήσῃ πλέον ἢ διπλάσιον τοῦ λεχθέντος πλάτους. φησὶ δ᾽ οὖν τὸ μὲν τῆς Ἰνδικῆς μέχρι τοῦ Ἰνδοῦ ποταμοῦ τὸ στενώτατον σταδίων μυρίων ἑξακισχιλίων (τὸ γὰρ ἐπὶ τὰ ἀκρωτήρια τεῖνον τρισχιλίοις εἶναι μεῖζον), τὸ δὲ ἔνθεν ἐπὶ Κασπίους πύλας μυρίων τετρακισχιλίων, εἶτ᾽ ἐπὶ τὸν Εὐφράτην μυρίων, ἐπὶ δὲ τὸν Νεῖλον ἀπὸ τοῦ Εὐφράτου πεντακισχιλίων, ἄλλους δὲ χιλίους καὶ τριακοσίους μέχρι Κανωβικοῦ στόματος, εἶτα μέχρι τῆς Καρχηδόνος μυρίους τρισχιλίους πεντακοσίους, εἶτα μέχρι στηλῶν ὀκτακισχιλίους τοὐλάχιστον· ὑπεραίρειν δὴ τῶν ἑπτὰ μυριάδων ὀκτακοσίους. δεῖν δὲ ἔτι προσθεῖναι τὸ ἐκτὸς Ἡρακλείων στηλῶν κύρτωμα τῆς Εὐρώπης, ἀντικείμενον μὲν τοῖς Ἴβηρσι προπεπτωκὸς δὲ πρὸς τὴν ἑσπέραν, οὐκ ἔλαττον σταδίων τρισχιλίων, καὶ τὰ ἀκρωτήρια τά τε ἄλλα καὶ τὸ τῶν Ὠστιδαίων, ὃ καλεῖται Κάβαιον, καὶ τὰς κατὰ τοῦτο νήσους, ὧν τὴν ἐσχάτην Οὐξισάμην φησὶ Πυθέας ἀπέχειν ἡμερῶν τριῶν πλοῦν. ταῦτα δ᾽ εἰπὼν τὰ τελευταῖα οὐδὲν πρὸς τὸ μῆκος συντείνοντα προσέθηκε τὰ περὶ τῶν ἀκρωτηρίων καὶ τῶν Ὠστιδαίων καὶ τῆς Οὐξισάμης καὶ ὧν φησι νήσων· ταῦτα γὰρ πάντα προσάρκτιά ἐστι καὶ Κελτικά, οὐκ Ἰβηρικά, μᾶλλον δὲ Πυθέου πλάσματα. προστίθησί τε τοῖς εἰρημένοις τοῦ μήκους διαστήμασιν ἄλλους σταδίους δισχιλίους μὲν πρὸς τῇ δύσει, δισχιλίους δὲ πρὸς τῇ ἀνατολῇ, ἵνα σώσῃ τὸ πλέον ἢ διπλάσιον τὸ μῆκος τοῦ πλάτους εἶναι.

Since he completely failed in relation to the breadth [sc of the inhabited world], it necessarily follows that he got it wrong as to its length. For later writers as well as the most accomplished earlier ones agree that its known length is more than twice its known breadth: I refer to the [distance] from the ends of India to the ends of Iberia and from Ethiopia up to the parallel of Ireland. But he defines his breadth as from the end of Ethiopia to the parallel through Thule, and so he extends his length more than is necessary, in order to make the length more than double the breadth. So he says that from the narrowest part of India to the river Indus is 16,000 stadia [1,820 miles, 2,910 km] (for up to its capes it extends another 3,000 stadia [340 miles, 545 km]). From there to the Caspian Gates is 14,000 stades [1,590 miles, 2,545 km]; to the Euphrates 10,000 stades [1,135 miles, 1,815 km], and to the Nile from the Euphrates 5,000 stades [570 miles, 910 km], and another 1,300 stades [150 miles, 240 km] to its Canobic mouth; then to Carthage 13,500 stades [1,535 miles, 2,455 km], and to Gibraltar at least 8,000 stades [910 miles, 1,455 km]. So that is 800 stades [90 miles, 145 km] more than 70,000 stades [7,955 miles, 12,730 km]. We must still add the bulge of Europe beyond the Pillars [Gibraltar] opposite the Iberians which projects to the west, not less than 3,000 stades [340 miles, 545 km], and the capes, especially that of the Ostidaioi called Cabaion, and the offshore islands, the furthest of which, Uxisamē, Pytheas says, is three days' sail away. Having noted these last places, which in no way extend the length, he has added the areas around the capes, both of the Ostidaioi and of Uxisamē, and the islands of which he speaks. But these are further north and belong to Keltikē, not Iberia; or rather they are fictions of Pytheas. He then adds to the above distances of length another 2,000 stades [230 miles] in the west and 2,000 in the east, in order to keep the length more than double the width.

This text continues from F25; Strabo turns to Eratosthenes' length of the *oikoumenē*, the habitable zone. The latter had the advantage over predecessors, who could only speculate;[1] he could access tangible information, from Pytheas for the north, and for the east from distances available after Alexander's conquests, of variable accuracy, as noted on F24. For Strabo, since he had wrongly believed Pytheas and extended the breadth up to Thule (F25), he need to extend the length, to maintain a ratio of over 2:1. The arithmetic is correct: 70,800+3,000+2,000+2,000 total 77,800 (stades; 8,840 miles, 14,145 km); Eratosthenes' width, as given in F24, is 38,000 stades (4,320 miles, 6,910 km).[2] Strabo gives his own dimensions at 2.5.6: 70,000 by less than 30,000 stades (7,955 miles × −3,410 miles, 7,955 × −5,455 km); cf on F33.

There are problems with the distances in modern terms, but Eratosthenes had to work with such information as was available. Since we know the shape of India, the 'narrowest part' and the 'capes' are difficult to understand, but reflect how it was envisaged: see on F27 with n 1. There are a number of locations for the Caspian Gates that suit our references to them; the context here suggests one of the two passes a little to the south or south-east of the Caspian Sea. We should note a detail about the distance between the most easterly branch of the Nile and its most westerly, the Canopic mouth (Canobic in Greek): in Strabo's MSS it is 1,500 stades, but editors consistently amend to 1,300 stades to make the arithmetic correct. But Strabo's mathematics was poor, as noted elsewhere, and 1,500 stades, about 170 miles, 270 km, could be what he wrote. Eratosthenes lived in Alexandria, and had access to local figures. It is some 145 miles, 230 km, as the crow flies; his distance, whether based on local roads or mariners' distances, would be rather more. The Atlantic coasts of Iberia and France seem to have been little understood. For Strabo, the Iberians lived on the Mediterranean coast of Spain, so 'opposite' them means further west: see Radt *ad loc*, though Roseman (1994) 124 noted Strabo's view of Spain and the river Tagus at 3.1.6, and suggested that 'opposite' means those living south of the river. 'What projects' should mean up to the Sacred Cape

1 Agathem 1: Democritos (c 460–c 370), and later Dicaiarchos, proposed 3:2; Eudoxos 2:1, which became the standard pattern for maps (Gem 16.3–4).
2 26,500 stades to the mouth of the Borysthenes plus 11,500 stades to the latitude through Thule.

(Cape St Vincent) of F34.[3] See also Strabo's view of the shape of Spain, Appendix 7 with maps 3a and 3b.

The coast from Penmarch northwards has many places that both locals and Pytheas would call a cape, but Cape Cabaion, Gabaion in Ptolemy *Geog* 2.8.1,2,5, is usually identified with Pointe du Raz. But it may well be Pointe de St Matthieu near Le Conquet (see Bianchetti (1998) 128); the 'offshore islands' are more easily identified with the string of islands running north-west between it and Uxisamē, Ushant (Axanthos in Pliny, F18), not the occasional islands further south closer to the Loire estuary.[4] It is not clear from Strabo's text whether Eratosthenes thought that Cabaion projected westwards to about the same extent as the Sacred Cape. In F37 he says that it projects quite far, though not as far as Pytheas says. As discussed in Appendix 7, it is also not clear how he thought of the cape in relation to Britain. The way in which Strabo jumps from the 'bulge', west of Gibraltar, to Brittany suggests either than he has considerably compressed how Eratosthenes described the area, based substantially on Pytheas, or Pytheas wrote so little about this stage of his voyage that Eratosthenes in turn could say little. Our basic sources for Pytheas are F34 to F37, and see the Coda pp 171–2. The whole must be read in the light of how Strabo thought of the orientation of the Iberian and French coasts, Appendix 7. As to Brittany, it is unclear whether Pytheas sailed to Ushant or just reported what he was told locally. From, say, Le Conquet, the harbour just north of Pointe St Mattheu, Ushant is 16 miles as the crow flies, well within a day's sail. Even from Pointe du Raz it is only 33 miles, 53 km. This suggests that the three days starts further south. For men familiar with the route, three days and nights across the open sea from Corbilo on the Loire estuary, F36, would be feasible (some 180 miles, 290 km); perhaps Pytheas learnt it there. If he went himself, his three days would need to begin further north. Much of the coast north of the Loire estuary is sinuous and some parts very rocky with dangerous reefs; he would keep close to the shore and make slow progress.[5] The Ost- tribe are in Brittany; see Appendix 5 for their spelling. Strabo closes with another anti-Pytheas comment, here using πλάσματα, *plasmata*, 'fiction'. In F24 and F25, he eliminated Thule (see on F25 for whether he thought it existed). Here, he could not eliminate northern Europe, but could assert that Pytheas' description was fiction; he does not state what were Pytheas' lies. What is not clear, at least from the way in which Strabo sets it out, is how Eratosthenes fitted in the two extra 2,000 stades at each end of his *oikoumenē*.

3 At F34, Strabo puts Cadiz to Cape St Vincent at 1,700 stades; as there noted, Cadiz is about 85–90 miles, 135–145 km, about 750 stades, from Gibraltar, so the two passages are not inconsistent.
4 As it happens, Pointe du Raz (4° 44′ 25″) is not further west than Pointe St-Mathieu (4° 46′ 26″), though Pytheas could not work to such limits.
5 The statue of the Virgin and Child known as Notre Dame des Naufragés, erected in 1904 at Pointe du Raz, is an eloquent witness to the dangers.

F27 Strabo 2.1.12

καὶ τὰ ἑξῆς δὲ πλήρη μεγάλων ἀποριῶν ἐστίν. ὅρα γάρ, εἰ τοῦτο μὲν μὴ κινοίη τις τὸ τὰ ἄκρα τῆς Ἰνδικῆς τὰ μεσημβρινὰ ἀνταίρειν τοῖς κατὰ Μερόην, μηδὲ τὸ διάστημα τὸ ἀπὸ Μερόης ἐπὶ τὸ στόμα τὸ κατὰ τὸ Βυζάντιον ὅτι ἐστὶ περὶ μυρίους σταδίους καὶ ὀκτακισχιλίους, ποιοίη δὲ τρισμυρίων τὸ ἀπὸ τῶν μεσημβρινῶν Ἰνδῶν μέχρι τῶν ὀρῶν, ὅσα ἂν συμβαίη ἄτοπα. τὸ πρῶτον μὲν γάρ, εἴπερ ὁ αὐτός ἐστι παράλληλος ὁ διὰ Βυζαντίου τῷ διὰ Μασσαλίας – καθάπερ εἴρηκεν Ἵππαρχος πιστεύσας Πυθέᾳ –, ὁ δ' αὐτὸς καὶ μεσημβρινός ἐστιν ὁ διὰ Βυζαντίου τῷ διὰ Βορυσθένους, ὅπερ καὶ αὐτὸ δοκιμάζει ὁ Ἵππαρχος, δοκιμάζει δὲ καὶ τὸ ἀπὸ Βυζαντίου διάστημα ἐπὶ τὸν Βορυσθένη σταδίους εἶναι τρισχιλίους ἑπτακοσίους, τοσοῦτοι ἂν εἶεν καὶ οἱ ἀπὸ Μασσαλίας ἐπὶ τὸν διὰ Βορυσθένους παράλληλον, ὅς γε διὰ τῆς Κελτικῆς παρωκεανίτιδος ἂν εἴη· τοσούτους γάρ πως διελθόντες συνάπτουσι τῷ Ὠκεανῷ . . .

The next statements [of Hipparchos] are also full of great difficulties. Look, if on the one hand one did not reject the southern capes of India rising opposite the areas around Meroë, and that Meroë to the mouth of the strait at Byzantium is about 18,000 stades [2,045 miles], but on the other hand make the distance from southern India to the mountains 30,000 stadia [3,410 miles], how many absurdities would follow. In the first place, if Byzantium is on the same latitude as Marseille (as Hipparchos says, trusting Pytheas), and that Byzantium is on the same longitude as the mouth of the Dnieper, which Hipparchos also proves, and if he also proves that the distance from Byzantium to the mouth of the Dnieper is 3,700 stadia [420 miles], then Marseille should be the same distance from the latitude of the mouth of the Dnieper; which latitude would also run through the coast of Keltikē, for going about this distance through Keltikē you reach the Ocean.

This text comes from where Strabo discusses Hipparchos' geography; it and F28 are from a section where he looks at Eratosthenes' geography in the light of that of Hipparchos. It mostly concerns eastern areas, his argument being that Hipparchos and his sources were wrong, because other authors have different distances and details, which he (Strabo) prefers; among Hipparchos' wrong sources was Pytheas the liar. For the east, both were right: once Alexander had opened up Persia and India, many new distances circulated with varying degrees of accuracy; see on F24 n 1. For India, the salient points are that (a) Dicaiarchos had established what became the standard latitude from Gibraltar through Rhodes to the Taurus mountains in Turkey and on the Himalayas (p 22); (b) but the Himalayas lie south of the Taurus range, and run from north-west to south-east. Eventually it became difficult to reconcile all the north-south distances depending on this latitude; (c) Eratosthenes stated that the southern capes of India (the mention of 'capes' must mean based on mariners' distances) were on the same latitude as Meroë; this was hopelessly wrong, but became an accepted standard;[1] (d) there were at least two distances recorded for the distance from the Himalayas to the southern tip of India, 20,000 and 30,000 stades (2,275, 3,410 miles; 3640, 5,455 km: Strabo 2.1.4); Strabo selected the latter, which strengthened his argument here, but overlooked the point that the former is not significantly different from his 18,000 stades for the same distance further west. Factually, the Pamirs, now in Tajikistan, are about 2,200 miles, roughly 20,000 stades, from the southern tip of India in a north-westerly direction; Srinagar, in the Himalayas more or less north of the tip, is some 1,800 miles, 2,880 km, distant, about 16,000 stades. The Pamirs are a little south of the latitude through Byzantium; Srinagar on about the same latitude as Beirut. However wrong we say Hipparchos and Strabo were, both were right, juggling with the material they had.

The references to Marseille and the coast of Keltikē, if they stood alone, would raise no difficulties. Strabo refers to Hipparchos putting Marseille on the same latitude as Byzantium, using Pytheas' *gnomon* reading for Marseille, elsewhere: F25, F31, F32, 2.4.3. If the latitude through the Dnieper is 3,700 stades, 420 miles, 670 km, further north, then at that distance north of Marseille it runs through Keltikē. He repeats this a couple of pages later, 2.1.16. In F31 he gives the distance as 3,800 stades, 430 miles, 690 km: see Appendix 7. It happens to be correct; Odessa, near the mouth of the Dnieper, is on about the same latitude as Nantes.[2] However, Strabo disagreed. Firstly, he cites Hipparchos as running the Dnieper latitude through Britain, not Keltikē, in F25 and F31. Secondly, after our text, he criticises the places on Hipparchos' Dnieper latitude further east, arguing that they are further south because they produce good wine, whereas the Dnieper area and the middle of Keltikē are too cold for this; the vines have to be wrapped up in winter (i.e. against frost). The problems with his conception of north-west Europe generally are discussed in Appendix 7.[3]

F28 Strabo 2.1.18

φησὶ δέ γε ὁ Ἵππαρχος κατὰ τὸν Βορυσθένη καὶ τὴν Κελτικὴν ἐν ὅλαις ταῖς θεριναῖς νυξὶ παραυγάζεσθαι τὸ φῶς τοῦ ἡλίου περιιστάμενον ἀπὸ τῆς δύσεως

1 Strabo 2.1.2. Meroë is nearly 17° N; the southern tip of India is about 8° N, a difference of some 600–620 miles, 960–990 km, over 5,000 stades. The southern tip is opposite the coast of Sri Lanka about halfway between the northern tip of the island and Colombo; Sri Lanka had been placed around 400 miles, 640 km, south of the Meroë latitude (see on F24).
2 Odessa, 46° 30′ N; Nantes, 47° 13′ N.
3 His complex criticism of Hipparchos' Dnieper latitude, 2.1.13–16, makes fascinating reading, e.g. in Jones' Loeb translation, and in the commentary by Roller (2018) 64–7. He concludes at 2.1.16 by repeating that Keltikē is 3,700 stades north of the Marseille-Byzantium latitude. Roseman (1994) 40–1 n 26 points out that some French wine-producing areas only go back to the early years of our era.

ἐπὶ τὴν ἀνατολήν, ταῖς δὲ χειμεριναῖς τροπαῖς τὸ πλεῖστον μετεωρίζεσθαι τὸν ἥλιον ἐπὶ πήχεις ἐννέα, ἐν δὲ τοῖς ἀπέχουσι τῆς Μασσαλίας ἑξακισχιλίοις καὶ τριακοσίοις, οὓς ἐκεῖνος μὲν ἔτι Κελτοὺς ὑπολαμβάνει, ἐγὼ δ᾽ οἶμαι Βρεττανοὺς εἶναι, βορειοτέρους τῆς Κελτικῆς σταδίοις δισχιλίοις πεντακοσίοις, πολὺ μᾶλλον τοῦτο συμβαίνειν· ἐν δὲ ταῖς χειμεριναῖς ἡμέραις ὁ ἥλιος μετεωρίζεται πήχεις ἕξ, τέτταρας δ᾽ ἐν τοῖς ἀπέχουσι Μασσαλίας ἐννακισχιλίους σταδίους καὶ ἑκατόν, ἐλάττους δὲ τῶν τριῶν ἐν τοῖς ἐπέκεινα, οἳ καὶ κατὰ τὸν ἡμέτερον λόγον πολὺ ἂν εἶεν ἀρκτικώτεροι τῆς Ἰέρνης. οὗτος δὲ Πυθέᾳ πιστεύων κατὰ τὰ νοτιώτερα τῆς Βρεττανικῆς τὴν οἴκησιν ταύτην τίθησι, καί φησιν εἶναι τὴν μακροτάτην ἐνταῦθα ἡμέραν ὡρῶν ἰσημερινῶν δέκα ἐννέα, ὀκτωκαίδεκα δὲ ὅπου τέτταρας ὁ ἥλιος μετεωρίζεται πήχεις, οὓς φησιν ἀπέχειν τῆς Μασσαλίας ἐννακισχιλίους καὶ ἑκατὸν σταδίους· ὥσθ᾽ οἱ νοτιώτατοι τῶν Βρεττανῶν βορειότεροι τούτων εἰσίν . . .

At any rate Hipparchos says that in all the summer nights around the mouth of the Dnieper and in Keltikē the light of the sun shines to some extent as it goes round from west to east; and at the winter solstices the sun rises at most around 9 cubits [18°]. Among those who are 6,300 (stades) [715 miles, 1,145 km] distant from Marseille, whom he understands to be Celts, but I think are Britons, who are 2,500 stades [285 miles, 455 km] north of Keltikē, this is even more the case; in winter days the sun rises 6 cubits [12°], and 4 cubits [8°] among those who are 9,100 stades [1,035 miles] from Marseille, and less than 3 cubits [6°] among those who are further away (who in my view would be much further north than Ireland). But trusting Pytheas, he puts this inhabited area further south than Britain, and says that the longest day there is 19 equal hours, and 18 hours where the sun rises only 4 cubits [8°]; he says that the people there are 9,100 stades [1,035 miles, 1,655 km] from Marseille; so that the most southerly of the Britons are more northerly than these people.

See the introduction to F27, including n 3, for the context, Strabo's criticisms of Hipparchos for criticising Eratosthenes. A page earlier, 2.1.17, he objected to Hipparchos placing Ariana and Bactria (between Persia and India) further north than Ireland; anywhere north of the latitude of Ireland was too cold to be habitable (F25, F31). He seeks to justify this with a complicated series of distances from the equator northwards, concluding with a reference to the Dnieper and Keltikē. Our text follows, attacking Hipparchos for northern Europe. It must be read in the light of how he envisaged Britain lay in relation to both Iberia and Keltikē: Appendix 7 and maps 3a and 3b.

Strabo makes use of data from Hipparchos' table or tables of latitudes and daylight, which are noted in Appendix 2 §§5–6. At 2.5.34, he objects to Hipparchos giving figures for the equator, because it is too hot to be habitable; at 2.5.43, just before F33, he says that regions too cold to be habitable do not concern a geographer. He does not systematically set out Hipparchos' table(s), but cites individual figures to suit his argument. In so doing, he converts Hipparchos' latitudes into stades. Hipparchos used Eratosthenes' circumference of the earth, 250,000 stades (pp xix–xx); a quadrant from the equator to the pole, 90°, was therefore 63,000 stades, and 1° = 700 stades. The table(s) were mostly for places on the standard longitude from the Cinnamon region south of Meroë to the Black Sea (F24), and further north into Scythia; this text shows that Hipparchos included locations in France and Britain. It makes for clarity to set out this part of the table(s) essentially as Strabo had it in front of him. As shown in Appendix 2 §6, Hipparchos could calculate figures for areas for which he had no readings, with fair accuracy; the 6,300 and 9,100 stades here, exact 7x and 13x multiples of 700 stades, are examples. The cubit is not the ancient unit of length, elbow to tip of finger, about 18″, but an astronomical unit for two of the 360° into which the circle was divided by Hipparchos' time.[1] Modern figures are noted in italics.

Marseille-Byzantium latitude, 15¼ hours (F32)
  Marseille 15 hrs 26 min; Istanbul 15 hrs 8 min

Mouth of the Dnieper latitude and Keltikē, 16 hours; 9 cubits (Strabo 2.5.42 and here)
  Odessa, 15 hrs 49 min, or Olbia, now Parutyne, Ukraine: 15 hrs 51 min, 20.1° = 10 cubits

9° north of Marseille-Byzantium, in Keltikē and Scythia, 17 hours; 6 cubits (Strabo 2.5.42 and here)
  9° = 6,300 stades = 715 miles, say Bremerhaven 17 hrs 3 min, 16.7° = 8.5 cubits

13° north of Marseille-Byzantium latitude, [location not stated] 18 hours; 4 cubits (here)
  13° = 9,100 stades = 1,035 miles, say Inverness 18 hrs 1 min, 12.8° = 6.5 cubits

[distance and location not stated] 19 hours, less than 3 cubits (here)
  Haroldswick, Unst, the northernmost Shetland, 19 hrs 10 min, 9.5° = 4.5 cubits

Modern figures must be read with care, given the inaccuracy of the Marseille-Byzantium latitude noted on F25. The table in Roseman (1994) 42–3 stresses the resulting differences for higher latitudes;[2] even so, Hipparchos' figures, at least for the summer solstice, are very close to modern ones.

Hipparchos' statement about light summer nights, which Strabo repeats at 2.5.42, is perhaps an incorrect inference by him; one has to reach the latitude of

---

1 Babylonian in origin, where it was 2½°, the Greeks made it 2°: Toomer (1984) 322 n 5; the 'finger' as an astronomical unit also noted in Toomer is amplified in Bowen and Todd (2004) 131 n 7. There is a question as to whether an astronomer called Hypsicles was the first to use the astronomical cubit: Dicks (1966) 27.
2 For 16 and 17 hours, compare Lyons, France, 16 hrs 2 min; Nottingham, 16 hrs 55 min.

St Petersburg, 59° 56', for white nights; it could not be accurate information from Greek settlements in the Black Sea such as Olbia. Hipparchos uses the rare word παραυγάζεσθαι, paraugasthai, 'to shine to some extent'. It aptly describes the half-light of short nights in the far north in high summer, and it is tempting to think that he found it in Pytheas, who used it to describe his own experience in the far north.[3]

Since he was writing well before the Roman conquest of Gaul, and possibly before the Rhine became known, or known with any accuracy, Hipparchos had little idea where Keltikē ended and Scythia began (cf p 10 and Note to Appendix 4). Even so, there is no problem about his putting the area 6,300 stades north of the Marseille latitude, with 17 hours of daylight, in Keltikē, at least if he thought that it was north-east and not due north of Marseille. Leeuwarden, though east of the Rhine, has 17 hours.[4] But Strabo seems to have thought of Keltikē as narrower that it is (so he would be correct to argue that 6,300 stades ran through Britain: see Appendix 7 p 164 for how Strabo envisaged the area). Note that it was not appreciated that Britons were Celts; Strabo did say that Britons had similar customs to Celts, though more barbaric (4.5.2; see on F38); so we cannot interpret Hipparchos' 6,300 stades from Marseille as referring to Celts in Britain.

The rest of his attack is a muddle. He wanted to argue that just as Hipparchos' figures for Ariana and Bactria were wrong, because he located them in the uninhabitable cold zone, although they are inhabited, so the 18 and 19 hour figures here were wrong, because 9,100 stades from Marseille took you into the cold zone: 5,000 plus 4,000 stades (F25) takes you to the limit of the habitable zone. An extra 100 stades is hardly 'much further north' than Ireland, but he was translating Hipparchos' 13° into stades, as just noted. If he had stopped there, his attack would be logical, given its basis.[5] But Pytheas is to blame, and Strabo either loses the thread of his argument or had a curious idea of the relative locations of Britain and Keltikē; his text cannot be cured by exchanging 'north' and 'south', as has been proposed: cf Appendix 7 and maps 3a and 3b.[6] The more difficult question for us is

---

3 Dicks (1960) 184 notes the suitability of the word but does not name Pytheas; Bianchetti (1998) 180–1 says that it could go back to him. Roller (2018) 68–9 positively proposes that it is Pytheas' word for arctic light, but does not comment on its use for the mouth of the Dnieper. At 123, on Strabo's later passage, 2.5.42, he says that the word might be local knowledge, noting that Eratosthenes' teacher Bion was born in Olbia; but he would know that there are no especially light nights in summer there.

4 In fact 16 hrs 58 min. Dicks (1960) 185 proposes that he thought of Keltikē extending to Jutland. Strictly wrong, because Jutland was not yet known (see on F17), but correct if we think of Holland and north Germany.

5 And we could then also suggest that Strabo knew of Pliny's source(s) for his Dnieper latitude in F20, which ran through Britain.

6 Jacob (1912) 149–50 sees no reason to change the text. Discussions of this passage agree that Strabo got it wrong: Dicks (1960) 190 on his F50; Roseman (1994) 43–4; Bianchetti (1998) 185–6; Radt (2002) 198. Ironically, the area from Calais to Dunkirk is north of a slice of southern Britain; but Strabo probably would not have known that.

what Pytheas said: was it geography or astronomy? It is more probably the former. Pytheas gave dimensions for Britain (F4), and described the Northern Isles (F19). Hipparchos would see that his 18 and 19 hour daylights well suited the north of Britain and the Northern Isles. Strabo would envisage areas with such daylight as outside the habitable zone; as noted in Appendix 7, the reference to the British diet in F38 indicates that Britain was all habitable. If it was astronomy, it would probably be a *gnomon* reading at the winter solstice, taken when overwintering in the Shetlands (Coda p 174), which Hipparchos could use as a cross-check against his own calculations for winter sun heights.

F29 Strabo 2.3.5

οὐ πολὺ οὖν ἀπολείπεται ταῦτα τῶν Πυθέου καὶ Εὐημέρου καὶ Ἀντιφάνους ψευσμάτων. ἀλλ' ἐκείνοις μὲν συγγνώμη, τοῦτ' αὐτὸ ἐπιτηδεύουσιν, ὥσπερ τοῖς θαυματοποιοῖς· τῷ δ' ἀποδεικτικῷ καὶ φιλοσόφῳ, σχεδὸν δέ τι καὶ περὶ πρωτείων ἀγωνιζομένῳ τίς ἂν συγγνοίη; ταῦτα μὲν οὖν οὐκ εὖ.

This does not fall far short of the fabrications of Pytheas, Euhemeros and Antiphanes. They could be pardoned, since fiction was their business, just as we pardon conjurors; but who could forgive Posidonios, master of demonstration and philosopher, a candidate for high honours? This he did not do well.

This fascinating text tells us nothing about Pytheas' book. Continuing his discussion and criticism of Eratosthenes (cf F24–F26) and Hipparchos (F27–F28), Strabo turns to Posidonios (p xxi). He has just related at length Posidonios' stories of various distant voyages (2.3.4–5, that by Phoenicians round 'Libya', i.e. Africa, for the pharaoh Necho, taken from Herodotos (pp 12–13); and another, also round Africa, by 'a certain magos', related by Heraclides of Pontos (b c 390 BC).[1] The third, set out at considerable length, describes voyages to India and various adventures of one Eudoxos of Cyzicos (not to be confused with the astronomer) in the second century. Strabo then says that Posidonios did not believe the first two for lack of independent evidence, but accepted the Eudoxos story, and was wrong to do so; he attacks Posidonios by bracketing him with other 'liars', naming two writers who wrote fiction, and Pytheas.

---

1 Necho's expedition is at Hdt 4.42.2–4. The Phoenicians took three years and reported that they had the sun on their right, which Herodotos disbelieved. Despite modern scepticism (see Asheri *et al* ad loc), it is hard to see how men who had not been in the southern hemisphere could have worked this out theoretically, at least at that time. The other voyage was said to have been sponsored by Gelon, tyrant of Gela in Sicily (early fifth century BC). Strabo's text is ambiguous as to whether Magos was the man's name, or whether he was a magos, i.e. a Persian official or savant whom Greeks variously thought of as performing religious duties or as experts in law and tradition (see, e.g., *OCD* sv).

Euhemeros (early third century)[2] wrote of a journey to the imaginary island of Panachaea;[3] Antiphanes of Berga, a town in Thrace, (date unknown, but probably about the same period) wrote what we would call short stories. It led to the adjective 'Bergean' meaning 'romancer' or 'liar' (so in F30).[4] While accepting Posidonios' learning in general, putting him on the same level as writers of fiction over these stories suggests that for all his width of reading, Strabo had limited breadth of judgment; as Radt puts it, 'the . . . irritating touch [of his style] is that, in disputing the field with other writers, he often loses himself in finicky fault-finding'.[5] Ironically, modern scholarship regards much of the Eudoxos story as credible, recognising that it reflects the use of the monsoon by the Ptolemies as an aid to sailing to India (see Roller (2018) 90–3).

F30 Strabo 2.4.1–2 = Polyb 34.5.1–13

(Strabo 2.4.1) [1] Πολύβιος δὲ τὴν Εὐρώπην χωρογραφῶν τοὺς μὲν ἀρχαίους ἐᾶν φησί, τοὺς δ' ἐκείνους ἐλέγχοντας ἐξετάζειν Δικαίαρχόν τε καὶ ᾿Ερατοσθένη, τὸν τελευταῖον πραγματευσάμενον περὶ γεωγραφίας, [2] καὶ Πυθέαν, ὑφ' οὗ παρακρουσθῆναι πολλούς, ὅλην μὲν τὴν Βρεττανικὴν ἐμβατὸν ἐπελθεῖν φάσκοντος, τὴν δὲ περίμετρον πλειόνων ἢ τεττάρων μυριάδων ἀποδόντος τῆς νήσου. [3] προσιστορήσαντος δὲ καὶ τὰ περὶ τῆς Θούλης καὶ τῶν τόπων ἐκείνων, ἐν οἷς οὔτε γῆ καθ᾽ αὑτὴν ὑπῆρχεν ἔτι οὔτε θάλαττα οὔτ' ἀήρ, ἀλλὰ σύγκριμά τι ἐκ τούτων πλεύμονι θαλαττίῳ ἐοικός, [4] ἐν ᾧ φησι τὴν γῆν καὶ τὴν θάλατταν αἰωρεῖσθαι καὶ τὰ σύμπαντα, καὶ τοῦτον ὡς ἂν δεσμὸν εἶναι τῶν ὅλων, μήτε πορευτὸν μήτε πλωτὸν ὑπάρχοντα· [5] τὸ μὲν οὖν τῷ πλεύμονι ἐοικὸς αὐτὸς ἑωρακέναι, τἆλλα δὲ λέγειν ἐξ ἀκοῆς. [6] ταῦτα μὲν τὰ τοῦ Πυθέου, καὶ διότι ἐπανελθὼν ἐνθένδε πᾶσαν ἐπέλθοι τὴν παρωκεανῖτιν τῆς Εὐρώπης ἀπὸ Γαδείρων ἕως Τανάϊδος.

(Strabo 2.4.2) [7] φησὶ δ' οὖν ὁ Πολύβιος ἄπιστον καὶ αὐτὸ τοῦτο, πῶς ἰδιώτῃ ἀνθρώπῳ καὶ πένητι τὰ τοσαῦτα διαστήματα πλωτὰ καὶ πορευτὰ γένοιτο. [8] τὸν δ' ᾿Ερατοσθένη διαπορήσαντα, εἰ χρὴ πιστεύειν τούτοις, ὅμως περί τε τῆς Βρεττανικῆς πεπιστευκέναι καὶ τῶν κατὰ Γάδειρα καὶ τὴν ᾿Ιβηρίαν· [9] πολὺ δέ φησι βέλτιον τῷ Μεσσηνίῳ πιστεύειν ἢ τούτῳ. ὁ μέν τοί γε εἰς μίαν χώραν τὴν

---

2 It might be a nom de plume, as *euhemeros* meant 'happy'.
3 His distant island of Panchaea had an idealistic society, with a temple to Zeus inside which an inscription described how the gods were mortals who had been deified because they had greatly benefitted the world in their lifetimes. This has given rise to 'euhemerism', the idea that real events lay behind myths.
4 The title of his series of tales was probably Ἄπιστα ('Unbelievable', Steph sv Βέργη): 'an untrustworthy joke of history and myth', Ps-Scymn 655. What little is known about him is most accessible in William Smith's 19th century *Dictionary of Greek and Roman Biography and Mythology*, available on the web (though his Polybius reference should be to 34.6.15).
5 Radt, *BNP sv* Strabo: 'Etwas störend ist . . . , daß er sich bei der Auseinandersetzung mit anderen manchmal in tüftelige Krittelei verliert'.

Παγχαίαν λέγει πλεῦσαι· ὁ δὲ καὶ μέχρι τῶν τοῦ κόσμου περάτων κατωπτευκέναι τὴν προσάρκτιον τῆς Εὐρώπης πᾶσαν, ἣν οὐδ' ἂν τῷ Ἑρμῇ πιστεύσαι τις λέγοντι. [10] Ἐρατοσθένη δὲ τὸν μὲν Εὐήμερον Βεργαῖον καλεῖν, Πυθέᾳ δὲ πιστεύειν, καὶ ταῦτα δὲ μηδὲ Δικαιάρχου πιστεύσαντος. [11] τὸ μὲν οὖν 'μηδὲ Δικαιάρχου πιστεύσαντος' γελοῖον, ὥσπερ ἐκείνῳ κανόνι χρήσασθαι προσῆκον, καθ' οὗ τοσούτους ἐλέγχους αὐτὸς προφέρεται. [12] Ἐρατοσθένους δὲ εἴρηται ἡ περὶ τὰ ἑσπέρια καὶ τὰ ἀρκτικὰ τῆς Εὐρώπης ἄγνοια. [13] ἀλλ' ἐκείνῳ μὲν καὶ Δικαιάρχῳ συγγνώμη, τοῖς μὴ κατιδοῦσι τοὺς τόπους ἐκείνους· Πολυβίῳ δὲ καὶ Ποσειδωνίῳ τίς ἂν συγγνοίη; . . .

[1] *Polybius, in his geography of Europe, says that he ignores the older writers, and looks at those who criticise them: Dicaiarchos, Eratosthenes, who has produced the most recent detailed book on geography, [2] and Pytheas, by whom many have been misled. Pytheas says that he visited the whole of Britain that was accessible, and reported that the whole coast-line was more than 40,000 stades [4,545 miles, 7,270 km]. [3] He also included his story about Thule and those regions in which there was neither land strictly speaking nor sea nor air, but a sort of mixture of these things like a jellyfish [4] in which he says that the earth and the sea and everything oscillate as if they were bound together, on which you could neither walk nor sail. [5] The thing like a jellyfish he saw for himself; the rest he reports from hearsay. [6] That is what is in Pytheas, and also that having returned from there he visited the whole coast of the Ocean from Gadeira [Cadiz] to Tanais [the Don].*

[7] *Polybios also says that it is incredible that a private individual, and a poor one, could have travelled such distances by sea and by land. [8] Although Eratosthenes was at a loss whether he should believe these things, yet he believed Pytheas on Britain, and the area around Cadiz and Iberia. [9] He [Polybios] says that it is better to believe the Messenian [Euhemeros] than Pytheas; he says that he sailed to one country, Panchaea. By contrast, he [Pytheas] says that he explored the whole of northern Europe as far as the ends of the world: something no one would believe, even if Hermes said it. [10] He [Polybios] says that Eratosthenes called Euhemeros a Bergaean, but believed Pytheas, even though not even Dicaiarchos believed him. [11] The 'not even Dicaiarchos believed him' is laughable, as if it were proper for him [Eratosthenes] to use as a standard the man about whom he himself makes so many criticisms. [12] The ignorance of Eratosthenes about the west and north of Europe has already been said. But while we may pardon Eratosthenes and Dicaiarchos, because they had not seen those regions with their own eyes, yet who could pardon Polybios and Posidonios?*

Strabo moves on from Posidonios (see on F29) to Polybios, c 201–c 120, born into an important aristocratic family in Megalopolis, Achaea, but sent as a young man as a hostage to Rome. He rapidly adapted to his new home, and became a friend and mentor to Cornelius Scipio Aemilianus, whom he accompanied on at least one and probably two or three of the latter's periods of military service in

Spain and Africa. He sailed into the Atlantic with a squadron provided by Scipio, mostly down the west coast of Africa but also some way up the coast of Portugal.[1] His Roman patriotism is clear from his *History*, which dealt with the rise of Rome and its domination of much of the Mediterranean world between 220 and 144. Within it, he included a section of geography, book 34, which is lost, but partly survives in quotations, as here, where the paragraphs are numbered as in editions of Polybios. Names in square brackets are added for clarity; the Greek is clear as to who is meant. §§1–10 are quotations or summaries of Polybios, though the 'many have been misled' in §2 is probably Strabo's comment; §§11–13 are Strabo's comments.

The starting point of any analysis is §7. It is thought that his dislike of Pytheas stemmed from a combination of his background, his new-found patriotism, and his literary ambitions. He wanted to be the first to describe the Atlantic, and to claim the glory of that for Rome. He needed to dismiss an upstart Greek who claimed to have been there already, and so argued that the man was too poor to have done it; his perspective was an aristocrat's innate disdain of his inferiors.[2] The patriotism aspect is argued by Dion (1965) 448–50; the dent to his personal pride not to be first to describe the area, aggravated by his social status, is well put at Walbank (1972) 126–7, so (1979) 205; Roseman (1994) 9–10. §1 is explicit that he had read Pytheas' book, but his references to what Pytheas wrote must be read in the light of his view of the man, and we should be alive to hostile exaggeration and sarcasm, and quoting selectively to prove his own agenda.

§1 Dicaiarchos, who helps to date Pytheas (see the Coda p 166), made major contributions to scientific geography; as noted at p 22, his latitude from Gibraltar to the Himalayas became a standard feature; cf F27 and F31. Eratosthenes is described as πραγματευσάμενον, *pragmateusamenon*, from a verb connoting taking time and trouble, here meaning writing in detail, as we see in F24–F26.

§2 Polybios presumably included Pytheas in his list of earlier geographers for the express purpose of disparaging him. The 'many' whom he has misled, whether Polybios' word or Strabo's comment, is an exaggeration. Dicaiarchos was not misled: according to §§9–10, he rejected Pytheas. Eratosthenes used Pytheas. Hipparchos also used him, but he was Polybios' contemporary, and probably not yet published, and not mentioned here. We know of no other geographer, at least for the north.[3]

---

1 For Polybios' travels beyond Gibraltar, see Walbank (1956) 3–5, (1972) 11; Polyb 34.15.7.

2 His 'poor man' comment should also be considered in the light of his observation at 12.27.4, that travel is expensive compared to reading about a place; see also p 133, noting that Polybios' journeys were all within a sailing season.

3 Though if it is Strabo's comment, it could include Hipparchos, 'misled' by Pytheas in F27 and F28, and see F31. Later, Artemidoros was not misled, F34.

There is a textual problem about Pytheas in 'the whole' of Britain. Strabo's MSS read ἐμβατὸν ἐπελθεῖν, *embaton epelthein*, as printed here. *Embaton* is an adjective meaning 'accessible'; *epelthein* means (a) 'come to, 'approach', or (b) 'go over', 'traverse' (it means to do so by sea in §5). Editors of both Strabo and Polybios usually amend *embaton* to ἐμβαδόν, *embadon*, an adverb meaning 'on foot'. That would mean that the mariner Pytheas left his boat and walked all over the island; one only has to say it to see that it is nonsense: it would take forever, and what happened to his crew? With either word, the sense is 'everywhere'. If he had done anything like that, we would expect far more details about Britain to survive. The itinerary and timetable in the Coda pp 170–7 allow for some short stopovers, and we can accept that his natural curiosity would encourage him to find out something about this newly discovered island. We may have lost some of his text on this, but his stay in Cornwall, F5, noting the diet, F38, learning some island names as he progressed, F19, and overwintering in the Shetlands, Coda p 174, would be enough for Polybios, with a good dose of sarcastic exaggeration, to embellish it all into a grand tour, going 'all over Britain'.[4] Polybios' 'more than 40,000' stades for the circumference of Britain correctly reflects Pytheas' actual 42,000 stades, F4, where its accuracy is discussed.

§§3–5 These paragraphs are central to the whole question of where Pytheas went and what he did in the far north, and need discussing in the light of our other Thule and frozen sea references: see Appendix 6. Here, it is convenient to note the creature translated 'jellyfish', πλεύμων θαλάττιος, *pleumōn thalattios* (here in the dative; literally 'lung of the sea'). The passages in the biological works of Theophrastos (F6.40) and Aristotle (*pleumon*, lungs, by itself, *PA* 581a 18)) show that it refers to what are popularly called jellyfish or medusae, though these names can refer to a number of different marine animals, which modern biologists distinguish, but Greeks could neither do so, nor understand their complete life cycles. An additional complication is that modern nomenclature may be based on a mistaken understanding of ancient names.[5] The relevant point for Pytheas is that he had to find a suitable analogy to describe what would be very strange and difficult to understand for a Mediterranean readership.

---

4 Roseman (1994) 126 points out that Britain was already covered with trackways; correct in itself, but it begs the question. Roller (2018) 95 argued that Pytheas spent a long time ('months') in Britain. Jacob (1912) 150 saw that walking all over Britain was nonsense, and suggested adding τὴν, *tēn*, 'the', before the adjective *embaton*. Neither words are common, but their meanings in the present context are clear.

5 The *pleumon* is also at Arist *HA* 581a 11, but in a passage that may not be by Aristotle himself. At Plato *Phileb* 21c, Socrates says that someone who has no power of thought lives like a '*pleumon* or such sea creatures as live in a shell', a very general comment mentioning two very different species and so not helpful. Generally, Thompson (1947) 203 see also 5–6; older translations of our passages have erroneously referred to other creatures, e.g. mollusc (Jowett for the Plato passage; Thompson *op cit* 189: most have shells) or sea cucumber (offered as an alternative in Peck's Loeb *PA*; Thompson *op cit* 38).

§6 This is an important but problematic sentence. The natural meaning of 'from Cadiz to the Don' is from west to east, through the Mediterranean to the Black Sea. But the 'coast' is παρωκεανῖτις, *parōkeanitis* (here in the accusative). It occurs 26 other times in Strabo, and always means a coast on the outer sea, the Ocean (*par(a)*,'along'). Indeed, he expressly compares the 'coast of our sea' (ἡ καθ' ἡμας παραλία, *hē kath' hēmas paralia*) with the *parōkeanitis*, the coast along the Ocean, at 3.4.16 and 17.3.24. Assuming it was Polybios' word here, and not Strabo summarising him, he too would have used it in the same sense: he mentions the Ocean elsewhere (3.33.15, 16.39.6), and cannot have intended the word to refer to the coast of the Mediterranean or Black Sea.[6] So Polybios says: having returned from 'there', i.e. the far north and the 'jellyfish' sea, Pytheas sailed along the Ocean coast from Cadiz to the Don. On the face of it, that is nonsense: when he got back from the far north, he had already sailed from Cadiz as far as Britain. We should treat Polybios' wording as hostile exaggeration. The Tanais, the Don, flows into the Black Sea, and no other text has a northern river of that name. But the Tanais was treated as the frontier between Europe and Asia, specifically Scythia, e.g. Strabo 11.7.4, and cf 17.3.24; and Polybios used it to bolster his attack: Pytheas claimed to have sailed all the way along the Ocean coast of Europe as far as Scythia. The fact that he probably did not get further east than around Rostock (Appendix 4 §11) is not a counter-argument: quite apart from general ignorance of the area, Polybios' attack is rhetoric, not geographic accuracy: Pytheas was not only a poor man but claimed to have penetrated the unknown north of Europe to its limits. We may reject an argument, taking into account what is said in F22, that since the short stretch from Cadiz to Gibraltar is on the Ocean, Polybius used 'parōkeanitis' very loosely, and Pytheas wrote a second book on the Mediterranean and Black Sea, called World Guide. Polybios is attacking Pytheas for being too poor to have made even one voyage, to the north and the limits of the known world.[7] As noted in F22 and in the Coda, p 177, all our references to Pytheas' work imply his having written just one book, about a voyage in the Ocean. Other possible views of this phrase are summarised in the Note below.

For §7, see above. §8: It is not clear whether this is Polybios or Strabo's comment, but it helps to confirm that Pytheas passed through Gibraltar (pp 17–19). On either view, it shows that Eratosthenes' *Geography* used Pytheas for the stated areas; it does not prove that Polybios was attacking the mathematics behind Eratosthenes' Thule latitude. Strabo was to have difficulty with Eratosthenes' Spain: see on F34, where Pytheas' details were only some of things which he alleged Eratosthenes got wrong, and cf F26 (dimensions) and F35 (general criticism).

---

6 The word does not occur elsewhere, except in lexicographers.
7 Roseman (1994) 50 does not propose exaggeration, but does suggests that Polybios used 'Tanais' to symbolise the limits of the world. The fact that *parōkeanitis* is the word used here is an additional argument against the point noted on F22, that Pytheas could have added the Mediterranean as a supplement to his book.

§§9–10 'The Messenian' is Euhemeros; he and Antiphanes of Berga were writers of fiction; as noted on F29, their names came to mean 'romancer' or 'liar'.[8] Hermes was a god whose attributes included Zeus' messenger and technological skill, but also trickery and theft. The Pytheas and Eratosthenes mentions add nothing to §8. The phrase 'as far as the ends of the world' may be Polybios' exaggeration, but does no more than repeat his 'Cadiz to Don' expression above, and confirms that Pytheas said something about northern Europe.[9] Polybios does not say on what point Dicaiarchos disbelieved Pytheas, if 'disbelieved' fairly reflects what Dicaiarchos actually wrote; and he did not realise the inconsistency between this and §2: if Dicaiarchos disbelieved Pytheas, it shows that he was not misled.

§§11–13 appear to be Strabo's own comments on the foregoing and his earlier reviews of Eratosthenes and Posidonios; as we have seen in F24–F28, he does not hesitate to criticise his predecessors with whom he disagrees. The final sentence in this text is inaccurate rhetoric. Polybios got no further north than Portugal (see above). Posidonios reached Cadiz, and his surviving references to the north (to Cimbri and Germans) are based on what he learnt in the Mediterranean.[10]

Note: other explanations for 'Cadiz to the Don'

Bianchetti (1998) 167–9 notes all the various explanations that have been offered: basically (i) Dion (1966) 200–2 imaginatively argued that the text uniquely offers a northern Tanais, a northern mirror image, marked as the Neman (now the frontier between Lithuania and Kaliningrad) on his map fig II p 207; this argument could equally apply if we substituted the Vistula. (ii) Dion (1977) 180–9 proposed that Pytheas was sent by Alexander to explore the north; it included Pytheas' having to try to find a route from the northern Ocean to the upper reaches of the Don. When Alexander reached the Jaxartes, he wondered if it was in fact these upper reaches. See further on this the Coda, p 167. (iii) Pytheas left his boat in the Baltic, travelled overland to the Sea of Azov and the mouth of the Don by old trade routes, and returned home, presumably on some merchant shipping; see Roller (2018) 96. This has the merits of explaining the mention of the Aeolian islands, F22, but we would then have expected his route to be described by as from Tanais to Marseille. In any case, there is no sense in any our references that Pytheas abandoned his boat. (iv) Diodoros 4.56.3 says that Timaios gave an account of the return of the Argonauts as sailing up the Don, hauling the boat overland to the Ocean, and returning home through Gibraltar into the Mediterranean, adding that Timaios found this route in earlier writers; it is suggested that this route was

---

8 Polybios reportedly also said that Eratosthenes surpassed Antiphanes for believing Pytheas: 34.6.15 = Const Porph *De Virt et Vit* 113 with Walbank (1979) 595.
9 In F25, Strabo says that Pytheas wrote about northern Europe, but does not say what (they were lies, he says); Polybios reports the same here. From what we can recover, see Appendix 4.
10 Roller's generalisation to the contrary (2018) 96, if intended to refer to northern Europe, overlooks the detailed discussion as to the extent and limits of Posidonios' travels in Kidd (1988) at 16–21, based on T19 E–K (Strabo 4.4.5), T25–6 E–K (§13 of the above text and Prisc Lyd 72.12); and at 307–15, on F67 E–K, Ath 4.151e–152f; 322–6 (F73 E–K, Ath 4.153e); and 947–8 (F227b E–K, Eustath *Hom Il* 13.6).

inspired by what Pytheas wrote. But it could just have well been inspired by the old trade routes, imaginatively woven into the Medea story. When Apollonios, a younger contemporary of Timaios, wrote his poem about the Argonauts (see on F22), he had them sailing home up the Po and down the Rhone; as good a route in the mythical world as any other. (v) Bianchetti (1998) 207 notes the suggestion that Pytheas wrote a second book, a *Periodos Gēs* as a description and map of the world, as to which see above.

F31 Strabo 2.5.7–8

[7] . . . πάλιν δ' ἀπὸ τῆς Ἀλεξανδρείας ἐπ' εὐθείας τῇ ῥύσει τοῦ Νείλου πάντες ὁμολογοῦσι τὸν ἐπὶ Ῥόδον πλοῦν, κἀντεῦθεν δὲ τὸν τῆς Καρίας παράπλουν καὶ Ἰωνίας μέχρι τῆς Τρωιάδος καὶ Βυζαντίου καὶ Βορυσθένους. λαβόντες οὖν τὰ διαστήματα γνώριμα καὶ πλεόμενα σκοποῦσι τὰ ὑπὲρ τοῦ Βορυσθένους ἐπ' εὐθείας ταύτῃ τῇ γραμμῇ μέχρι τίνος οἰκήσιμά ἐστι, καὶ τί περατοῖ τὰ προσάρκτια μέρη τῆς οἰκουμένης. οἰκοῦσι δ' ὑπὲρ τοῦ Βορυσθένους ὕστατοι τῶν γνωρίμων Σκυθῶν Ῥωξολανοί, νοτιώτεροι ὄντες τῶν ὑπὲρ τῆς Βρεττανικῆς ἐσχάτων γνωριζομένων· ἤδη δὲ τἀπέκεινα διὰ ψῦχος ἀοίκητά ἐστι. νοτιώτεροι δὲ τούτων καὶ οἱ ὑπὲρ τῆς Μαιώτιδος Σαυρομάται καὶ Σκύθαι μέχρι τῶν ἑῴων Σκυθῶν.

[8] ὁ μὲν οὖν Μασσαλιώτης Πυθέας τὰ περὶ Θούλην τὴν βορειοτάτην τῶν Βρεττανίδων ὕστατα λέγει, παρ' οἷς ὁ αὐτός ἐστι τῷ ἀρκτικῷ ὁ θερινὸς τροπικὸς κύκλος· παρὰ δὲ τῶν ἄλλων οὐδὲν ἱστορῶ, οὔθ' ὅτι Θούλη νῆσός ἐστί τις, οὔτ' εἰ τὰ μέχρι δεῦρο οἰκήσιμά ἐστιν ὅπου ὁ θερινὸς τροπικὸς ἀρκτικὸς γίνεται. νομίζω δὲ πολὺ εἶναι νοτιώτερον τοῦτο τὸ τῆς οἰκουμένης πέρας τὸ προσάρκτιον· οἱ γὰρ νῦν ἱστοροῦντες περαιτέρω τῆς Ἰέρνης οὐδὲν ἔχουσι λέγειν, ἣ πρὸς ἄρκτον πρόκειται τῆς Βρεττανικῆς πλησίον, ἀγρίων τελέως ἀνθρώπων καὶ κακῶς οἰκούντων διὰ ψῦχος, ὥστ' ἐνταῦθα νομίζω τὸ πέρας εἶναι θετέον. τοῦ δὲ παραλλήλου τοῦ διὰ Βυζαντίου τοῦ διὰ Μασσαλίας πως ἰόντος, ὥς φησιν Ἵππαρχος πιστεύσας Πυθέᾳ (φησὶ γὰρ ἐν Βυζαντίῳ τὸν αὐτὸν εἶναι λόγον τοῦ γνώμονος πρὸς τὴν σκιάν, ὃν εἶπεν ὁ Πυθέας ἐν Μασσαλίᾳ), τοῦ δὲ διὰ Βορυσθένους ἀπὸ τούτου διέχοντος περὶ τρισχιλίους καὶ ὀκτακοσίους, εἴη ἂν ἐκ τοῦ διαστήματος τοῦ ἀπὸ Μασσαλίας ἐπὶ τὴν Βρεττανικὴν ἐνταῦθά που πίπτων ὁ διὰ τοῦ Βορυσθένους κύκλος. πολλαχοῦ δὲ παρακρουόμενος τοὺς ἀνθρώπους ὁ Πυθέας κἀνταῦθά που διέψευσται. τὸ μὲν γὰρ τὴν ἀπὸ Στηλῶν γραμμὴν ἐπὶ τοὺς περὶ τὸν Πορθμὸν καὶ Ἀθήνας καὶ Ῥόδον τόπους ἐπὶ τοῦ αὐτοῦ παραλλήλου κεῖσθαι ὡμολόγηται παρὰ πολλῶν· ὁμολογεῖται δέ, ὅτι καὶ διὰ μέσου πως τοῦ πελάγους ἐστὶν ἡ ἀπὸ Στηλῶν ἐπὶ τὸν Πορθμόν. οἱ δὲ πλέοντες τὸ μέγιστον δίαρμα ἀπὸ τῆς Κελτικῆς ἐπὶ τὴν Λιβύην εἶναι τὸ ἀπὸ τοῦ Γαλατικοῦ Κόλπου σταδίων πεντακισχιλίων, τοῦτο δ' εἶναι καὶ τὸ μέγιστον πλάτος τοῦ πελάγους, ὥστ' εἴη ἂν τὸ ἀπὸ τῆς λεχθείσης γραμμῆς ἐπὶ τὸν μυχὸν τοῦ Κόλπου σταδίων δισχιλίων πεντακοσίων, ἐπὶ δὲ Μασσαλίαν ἐλαττόνων· νοτιωτέρα γάρ ἐστιν ἡ Μασσαλία τοῦ μυχοῦ τοῦ Κόλπου. τὸ δέ γε ἀπὸ τῆς Ῥοδίας ἐπὶ τὸ Βυζάντιόν ἐστι τετρακισχιλίων που καὶ ἐννακοσίων σταδίων, ὥστε πολὺ ἀρκτικώτερος ἂν εἴη ὁ διὰ Βυζαντίου τοῦ διὰ Μασσαλίας. τὸ δ' ἐκεῖθεν ἐπὶ τὴν Βρεττανικὴν δύναται συμφωνεῖν τῷ ἀπὸ Βυζαντίου ἐπὶ Βορυσθένη· τὸ δ' ἐκεῖθεν ἐπὶ τὴν Ἰέρνην οὐκέτι γνώριμον πόσον

ἄν τις θείη, οὐδ’ εἰ περαιτέρω ἔτι οἰκήσιμά ἐστιν, οὐδὲ δεῖ φροντίζειν τοῖς ἐπάνω λεχθεῖσι προσέχοντας.

[7] *Again, all [geographers] agree that sailing from Alexandria to Rhodes is in a straight line with the course of the Nile, and from there sailing along Caria and Ionia to the Troad, Byzantium, and the mouth of the Dnieper. So taking the distances by sea that are known, they look at the regions beyond the Dnieper on the direct continuation of that line as to how far they are inhabitable, and what is the limit of the northern parts of the inhabitable zone. The most remote of the Scythians who live beyond the Dnieper that are known are the Roxolani, although they are further south than the most distant Britons who are known. The regions beyond there are uninhabitable because of the cold. Further to the south are the Sauromatians who live beyond the Sea of Azov, as well as the Scythians as far as the Eastern Scythians.*

[8] *Now Pytheas of Marseille says that the regions around Thule, the most northerly of the British isles, are the furthest [habitable regions], where the summer tropic is the same as the arctic circle. But from other [writers] I can record nothing, either that there is an island called Thule, or if the area up to there is habitable, where the summer tropic becomes the arctic [circle]. I think that the northern limit of the habitable zone is much further south. For those reporting on what is beyond Ireland, which lies near to and north of Britain, have nothing to say; it is inhabited by men who are complete savages and lead a miserable life because of the cold. So I think that is where the limit the habitable zone is to be placed. But if the latitude of Byzantium is more or less that of Marseille, as Hipparchos says trusting Pytheas (for he says that the shadow of the gnomon at Byzantium has the same proportion as that which Pytheas gave for Marseille), and if the mouth of the Dnieper is about 3,800 [stades] [430 miles, 690 km] from there, then with the distance from Marseille to Britain the latitude through the mouth of the Dnieper would pass somewhere in Britain. But Pytheas, who misleads people in many places, here also is somewhat mistaken. For it is agreed by many that a line from Gibraltar through the places around the Straits [of Sicily], Athens and Rhodes, lies on the same parallel; and it is agreed that the part of the line from the Pillars to the Straits runs more or less through the middle of the sea. Most people say that the longest passage from Keltikē to Libya [Africa] is that from the Galatian Gulf [Gulf of Lion], 5,000 stadia [570 miles, 910 km], and that it is also the greatest width of the [Mediterranean] sea. It follows that the distance from this line to the head of the gulf would be 2,500 stadia [285 miles, 455 km], less to Marseille; for Marseille is further south than the head of the gulf. But from Rhodes to Byzantium is about 4,900 stadia [555 miles, 890 km], and therefore the latitude through Byzantium would be much further north than that through Marseille. The distance from there [Marseille] to Britain may agree with that from Byzantium to the mouth of the Dnieper; but it is not yet known how much would one should add for the distance from Britain to Ireland, nor whether beyond that there are still habitable regions, nor is it necessary to consider adding to what has been said above.*

This text, like F32 and F33, comes from the final part of book 2. Having agreed with or criticised predecessors, Strabo now states his own views. As elsewhere, the important word is οἰκουμένη, *oikoumenē*, the habitable zone. Just before this text he gives his own figures for the *oikoumenē*, 70,000 by 30,000 stades (7,955 by 3,410 miles, 12,730 by 5,455 km: Strabo 2.5.6); see further Roller (2018) cited in F32 n 1. He now identifies its northern limit, north of the Black Sea (§7) and Ireland (§8).

§7 is sensible. The preceding text refers to the long-established longitude along the Nile, certainly in Eratosthenes, F24–F25. Rhodes is not mentioned there; even if Eratosthenes did not include it, later writers did: it was a major hub for Mediterranean traffic, and using Rhodes to help define the longitude from Alexandria to the Dardanelles and Byzantium was more precise.[1] In F24 Strabo uses the technical term *mesēmbrinos* ('midday'; cf our 'meridian', for longitude); here he has the more prosaic 'straight line'. Given his Stoic beliefs about habitable and inhabitable zones, his logic is impeccable: go north from the Black Sea to where the Roxolani live, and you reach the northern limit of the habitable zone. A page or so further on, 2.5.9, he puts this as 4,000 stades (455 miles, 730 km) north of the Dnieper latitude (as he had in F25). Although for Strabo it was a calculated figure on the basis of his beliefs about the limit of the *oikoumenē*, the figure happens to coincide with archaeology, which suggests that the Roxolani lived up to about latitude 52°, roughly the boundary between Ukraine and Belarus, or Pinsk, which is about 480 miles, 770 km, from the mouth of the Dnieper as the crow flies. Other Scythian tribes, he says, live further south; Greek knowledge of the tribes north of the Black Sea is a topic in its own right.[2] His explanation for the Roxolani being further south than the north of Britain follows in §8.

§8 As with §7, given his standpoint this is logical; whether his argument is sound is another matter. He is ambiguous as to whether Thule exists or not. Here, he says that others say nothing about it. But in F24 and F25 he complains about Eratosthenes' noting it, and in F38 he accepts that others do mention it. As noted on F38, information about Britain would come from the south, areas on the English Channel, where little if anything would be known about the Northern Isles or what lay beyond. Strabo confirms that the summer tropic and arctic circle coinciding on Thule, as recorded

---

1 Moreover, for geographers Rhodes was a sort of central point, because it also lay on Dicaiarchos' latitude from Gibraltar to the Himalayas, mentioned below.
2 The Roxolani are thought to have been a tribe within the Sauromatians (or Sarmatians). Greeks knew about Scyths from trade contacts; they controlled some of the silk road routes. They had very many tribal names for them or groups within them, as can be seen from Strabo 7.3.3–19, dealing (broadly speaking) with parts of Eastern Europe up to the Ukraine, including the areas touched on here. But neither he nor others had much specific information as to where they variously lived; also, there had been migrations westward since Herodotos had made his enquiries about them as recorded in his book 4, 1–36. There are helpful discussions in *BNP* svv Rhoxolani, Sarmatae, and Scythae, esp part 2, History, and the map at the beginning.

in F2, is from Pytheas; and notes that Pytheas treated it as a British island. That was geography; there is no implication of a political connection with other British islands or the mainland. Despite its six days' distance, F15 and F24, Pytheas had no reason to think otherwise. As in F25, Ireland is at the limited of the *oikoumenē* in the north-west, and he speaks of its cold and its savage inhabitants at 2.1.13 (just after F27) and in 2.5.8 and 4.5.4: see p 8 n 5 and Appendix 7. He reverts to its location further down.

He now has a confused argument which vacillates between impugning Pytheas' *gnomon* reading for Marseille (see on F25) and then conceding that it may be right. This is reflected in his rebuke, which is mild compared to his usual 'liar': Pytheas που διαψεύσται, *pou diapseustai*, 'is somewhat mistaken'. There is a tex-tual problem; with one exception the MSS have πανταχοῦ πολλαχοῦ, *pantachou pollachou*, 'everywhere in many places', which is illogical even for Strabo, and looks like dittography. I print *pollachou*, 'in many places': Pytheas' geography is usually wrong (Strabo had a grudging respect for his astronomy, F38, F39). As in F25, he notes that the two *gnomon* readings for Marseille and Byzantium are not quite the same; I have translated his word, πως, *pōs*, connoting somewhat or about, 'more or less' (the same). Except here, Strabo accepts the latitude as cor-rect; for that, and the 3,800-stade distance to the Dnieper latitude, and its running through Britain, see Appendix 7.

It is ironic that Strabo's reasoning was correct on the material he uses, and hope-lessly wrong in fact. He starts with Dicaiarchos' standard latitude from Gibraltar through Rhodes to the mountains of India, which has been noted at p 22 and on F27. The western part is reasonable by ancient standards: the latitude of Gibraltar, 36° 08′ N, runs through Rhodes about one-third of the way up the island; at F32 he says that it runs through the middle, a difference of some 7 miles, 11 km, irrel-evant by the then standards of accuracy. The problem with its eastern end is noted on F27. Strabo then says that the Gulf of Lion is 5,000 stades wide, and Dicaiar-chos' latitude bisects it. So there are 2,500 stades from that north to the head of the gulf, a little less to Marseille further south. Thus Marseille is a little less than 2,500 stades north of Rhodes, while Byzantium is 4,900 stades north. In fact, the latitude through Gibraltar does not bisect the Gulf of Lion; it runs through parts of northern Algeria and Tunisia, and does not even touch the sea anywhere oppo-site the south coast of France. That the Gulf is closer to 4,000 than 5,000 stades wide, and that it is not the 'widest point', are minor errors by comparison.[3] It is also ironic that Marseille is in fact slightly north of Byzantium: see on F25. Hav-ing proved his point, Strabo becomes uncomfortable: he realises that the distance from the Marseille to Britain is the same as that from Byzantium to the Dnieper. For him, it upsets his distance from Britain to the limit of the *oikoumenē*, Ireland,

---

3 The Gulf is about 460 miles, 735 km, as the crow flies, and Marseille some 20 miles, 32 km, south of its head. The longest stretch is from the heel of Italy to the head of the bay south-west of Benghazi, about 550 miles, 880 km.

already given as 4,000 stades at F25, the same as the distance to the Roxolani further east given here, and 'to the northern regions' at 2.5.9.

The final words of our text are clarified by what Strabo next writes. He says that distant areas are of no interest to 'leaders', governments in our terms, if their peoples do not harm Rome; for this reason, Romans, meaning Caesar, scorned to hold Britain, although they could have done; cf on F33. Roller (2018) 105 points out that Strabo was writing after Caesar's failure in Britain, so it would be appropriate to make it appear an island of little consequence.

F32 Strabo 2.5.41

ἐν δὲ τοῖς περὶ τὸ Βυζάντιον ἡ μεγίστη ἡμέρα ὡρῶν ἐστιν ἰσημερινῶν δέκα πέντε καὶ τετάρτου, ὁ δὲ γνώμων πρὸς τὴν σκιὰν λόγον ἔχει ἐν τῇ θερινῇ τροπῇ ὃν τὰ ἑκατὸν εἴκοσι πρὸς τετταράκοντα δύο λείποντα πέμπτῳ. ἀπέχουσι δ' οἱ τόποι οὗτοι τοῦ διὰ μέσης τῆς Ῥοδίας περὶ τετρακισχιλίους καὶ ἐννακοσίους, τοῦ δ' ἰσημερινοῦ ὡς τρισμυρίους τριακοσίους . . .

In the area of Byzantium the longest day has 15¼ equinoctial hours, and the ratio of a gnomon to the shadow at the summer solstice is 120 to 42 minus one fifth [600:209]. These regions are about 4,900 stadia [555 miles, 890 km] distant from the latitude of the centre of Rhodes and about 30,300 stadia [3445 miles, 5,510 km] from the equator.

This text was included by Mette: his F6c combined it with F31. Bianchetti and Roseman omitted it. However, it touches on Pytheas' *gnomon* reading for Marseille (F25), and illustrates the context for F33. It is extracted from Hipparchos' table(s) of latitude; it came next to the calculated figure of 15½ hours for the middle of the Black Sea (Appendix 2 §§5–7: the 15½ hour figure, n 5; cf on F28; for 'equinoctial hours', see on F7). There are two points to note about Hipparchos' *gnomon* reading. Firstly, it illustrates the limits of accuracy achievable. To measure the shadow, you need two rods: one held vertically and one to measure the shadow, as detailed in Appendix 2 §3. The larger number, 120, must reflect the length of Hipparchos' *gnomon* (in summer, the shadow must be the lower number). The basic Greek measure of length was the δάκτυλος, *daktylos*, 'finger', 0.76″, so 120 *daktyloi* corresponds to about 7′7″; that would be the known length of his vertical rod. If the other rod was marked in *daktyloi*, he would find that the shadow fell one-fifth short of the 42 mark. The limits of accuracy can be easily illustrated: tan 600/209 (see Appendix 2 §8) is 70.8°; the modern figure for the height of the sun at the summer solstice in Istanbul is 72.4°. Secondly, we should not assume that this was the reading for Byzantium which Hipparchos used when seeing that Pytheas' figure for Marseille enabled him to place both places on the same latitude (F25, F28, F31). This is not because of the slight differences in readings indicated in F25 and F31. Common sense suggests that Pytheas' voyage started in spring: see the timetable proposed in the Coda, pp 170–4, and it is likely that his *gnomon* reading was taken at the spring equinox just before leaving. If Hipparchos compared summer solstice readings, we have to assume that Pytheas took such a reading independently of his voyage but chose to record it in his book.[4]

It is typical of Strabo's carelessness with figures that the distances here differ from what he has elsewhere. See on F24 for a commentary on Eratosthenes' longitude, including that from the equator to Syene is 16,800 stades, plus 5,000 stades to Alexandria, 21,800. As reported, the next stages were the Hellespont and the Dnieper, 8,100 and 5,000 stades respectively: total 34,000 stades. Elsewhere Strabo offers 3,700 and 3,800 stades from Byzantium north to the Dnieper (F27, F31), which would give 31,200 or 31,100 stades from the equator to Byzantium. However, it became more usual to define the longitude via Rhodes, for which various distances were known: about 4,900 stades, F31 and here; 4,900 and 5,000 stades, Strabo 2.4.3; 4,000 and 5,000 stades, 2.5.24 (p xvii). From Rhodes to Byzantium is 5,000 stades, 2.4.3, and 4,900 stades here. But it is hard to see how he calculated his 30,300 stades. Assuming that he used the 21,800 stades figure for the equator to Alexandria, the stages to Rhodes and then Byzantium here would total only 8,500 stades.[5]

F33 Strabo 2.5.43

... τοῦ γὰρ ἡλίου καθ' ὅλην τὴν τοῦ κόσμου περιστροφὴν ὑπὲρ γῆς φερομένου, δῆλον ὅτι καὶ ἡ σκιὰ κύκλῳ περιενεχθήσεται περὶ τὸν γνώμονα. καθ' ὃ δὴ καὶ περισκίους αὐτοὺς ἐκάλεσεν, οὐδὲν ὄντας πρὸς τὴν γεωγραφίαν· οὐ γάρ ἐστιν οἰκήσιμα ταῦτα τὰ μέρη διὰ ψῦχος, ὥσπερ ἐν τοῖς πρὸς Πυθέαν λόγοις εἰρήκαμεν ...

For since the sun moves above the earth throughout the whole revolution of the universe, it is clear that the shadow will rotate in a circle round the gnomon. It was for this reason that he [Posidonios] called them 'periscians', although they do not exist as far as geography is concerned; for these regions are uninhabitable on account of the cold, as I have already said in passages about Pytheas.

This text comes a few pages after F32, towards the end of Strabo's own view of the world. Beginning at 2.5.35, he describes the habitable world in terms of the length of the longest day, starting from the south, adding various other details such as what stars are visible and *gnomon* readings (Appendix 2 §3), basically drawing on Hipparchos' table(s) (Appendix 2 §§5–7; cf on F28). He now again mentions what Posidonios said about different zones, which he had discussed at

---

4 However long Pytheas was away, he would be likely to leave in the spring and return in the autumn (Coda pp 170–7).
5 Roller (2018) 103–9, (2010) 146–7, 151–4 has useful discussions of these figures; he argues that the 70,000 x nearly 30,000 for Strabo's dimensions of the *oikoumenē* go back to Eratosthenes, though he accepts that the figures are not always consistent, and it is clear from F24 to F26 that Strabo does not accept Eratosthenes' figures.

2.2.3, in particular the latter's thinking about the direction of the sun's shadow at different latitudes. Posidonios invented three words: 'amphiscians', literally 'both-way shadowers', those living within the tropics, whose shadows might fall either south or north; 'heteroscians', 'either-way shadowers', those in temperate zones, whose shadows were always to the south in the Greeks' own temperate zone and always to the north for those in the corresponding zone south of the equator; and 'periscians', 'roundabout shadowers', those living above the arctic and antarctic circles, whose shadows would fall both south and north during the day.[1] Strabo had to accept that there could be amphiscians, because the Tropic of Cancer ran through Syene, and the Land of Cinnamon, the southern limit of the habitable zone, was further south (F24); but clearly no one could live so far north as to be periscians. Pytheas lied about the north, as Strabo never tires of saying. That such a region was of no interest to geographers is also just after F31, as there noted. After this text, Strabo concludes book 2 with a few words to the effect that he need say nothing further about this uninhabitable region.

F34 Strabo 3.2.11

. . . ἐοίκασι δ' οἱ παλαιοὶ καλεῖν τὸν Βαῖτιν Ταρτησσόν, τὰ δὲ Γάδειρα καὶ τὰς πρὸς αὐτὴν νήσους Ἐρύθειαν· διόπερ οὕτως εἰπεῖν ὑπολαμβάνουσι Στησίχορον περὶ τοῦ Γηρυόνος βουκόλου, διότι γεννηθείη

> σχεδὸν ἀντιπέρας κλεινᾶς Ἐρυθείας
> Ταρτησσοῦ ποταμοῦ παρὰ παγὰς ἀπείρονας ἀργυρορίζους,
> ἐν κευθμῶνι πέτρας.

δυεῖν δὲ οὐσῶν ἐκβολῶν τοῦ ποταμοῦ πόλιν ἐν τῷ μεταξὺ χώρῳ κατοικεῖσθαι πρότερόν φασιν ἣν καλεῖσθαι Ταρτησσόν, ὁμώνυμον τῷ ποταμῷ, καὶ τὴν χώραν Ταρτησσίδα ἣν νῦν Τουρδοῦλοι νέμονται. καὶ Ἐρατοσθένης δὲ τὴν συνεχῆ τῇ Κάλπῃ Ταρτησσίδα καλεῖσθαί φησι, καὶ ''Ἐρύθειαν'' νῆσον εὐδαίμονα. πρὸς ὃν Ἀρτεμίδωρος ἀντιλέγων καὶ ταῦτα ψευδῶς λέγεσθαί φησιν ὑπ' αὐτοῦ, καθάπερ καὶ τὸ ἀπὸ Γαδείρων ἐπὶ τὸ Ἱερὸν Ἀκρωτήριον διάστημα ἀπέχειν ἡμερῶν πέντε πλοῦν, οὐ πλειόνων ὄντων ἢ χιλίων καὶ ἑπτακοσίων σταδίων, καὶ τὸ τὰς ἀμπώτεις μέχρι δεῦρο περατοῦσθαι ἀντὶ τοῦ κύκλῳ περὶ πᾶσαν τὴν οἰκουμένην συμβαίνειν, καὶ τὸ τὰ προσαρκτικὰ μέρη τῆς Ἰβηρίας εὐπαροδώτερα εἶναι πρὸς τὴν Κελτικὴν ἢ κατὰ τὸν Ὠκεανὸν πλέουσι, καὶ ὅσα δὴ ἄλλα εἴρηκε Πυθέᾳ πιστεύσας δι' ἀλαζονείαν.

Our ancestors seem to have called the river Baetis 'Tartessos'; and Gadeira [Cadiz] and the nearby islands 'Erytheia'; because this is how people understand Stesichoros when he speaks of Geryon the cowherd, since he was born

---

1 Posidonios' amphi-, hetero- and periscians fall within his general views of the world; see Kidd (1989) 229–31 on F49 E–K = Strabo 2.2.3, and *id* 742–7 on F208 E–K = this text. They are referred to by Cleomedes I.4.132–46, cf 225–31; and Ptolemy in various passages in his *Syntaxis* book 1, but they are not an idea that seems to have been developed further.

*almost opposite famous Erytheia / beside the unlimited silver-rooted
springs of the Tartessos river / in a cavern in a cliff.*

As the river had two mouths, they say that a city was formerly founded between
them called Tartessos after the river; and the land where the Turduli now live
was called 'Tartessis'. Eratosthenes says that the area next to Calpē [Gibraltar]
is called 'Tartessis', and Erytheia 'the prosperous isle'. Artemidoros disagrees
and says that these things are false statements by him, just like his saying that
the distance from Gades to the Sacred Cape is five days' sailing, although it is
not more than 1,700 stades [195 miles]; that tides come to an end at the Sacred
Cape, although they occur in a circle all round the inhabited world; and that the
northern parts of Iberia offer easier access to Keltikē than if you sail round by the
Ocean; and all other matters which he said relying on Pytheas on account of the
latter's falsehoods.

Strabo's book 3 describes Spain, from which this text and F35 come. After a
general overview, and describing its shape, for which see Appendix 7, he starts
at the south-west; his first geographical detail, 3.1.4, is the Sacred Cape also
mentioned here. Tartessos was the name by which Greeks knew the south of
Spain west of Gibraltar: p 10, and in more detail pp 14–16. As there discussed,
it was rich in both silver and tin, and Greeks traded profitably there. Herodotos
1.163 and 4.152 refers to Phocaea and Colaios of Samos, but others must have
gone there, and it is generally assumed that this included Massalia. We can only
assess its extent from Strabo's description of the position in his time; that would
place it as the land between the rivers Anas (Guadiana, now the frontier between
Spain and Portugal) in the west, and Baetis (Guadalquivir), and possibly further east to Gibraltar, as discussed below. It was possibly the Tarshish of the
bible.[1] Greeks knew of a *polis* called either Ἐλιβύργη, Elibyrgē, per Hecataios
of Miletos (c 500), *FGrH* 1 F38 = Steph Byz sv), or Ἴβυλλα, Ibylla (Herodianus
De Pros Cath 3.1.255, Steph Byz sv); probably the same place, the consonants
-b- and -l- in an unfamiliar foreign name being reversed in pronunciation. It
has been identified with Alcalá del Río, some 12 miles north of Seville (Barrington 26), but it could equally be the port.[2] The mouth of the Guadalquivir has
changed since antiquity, so it is not surprising that no port or trading post has
been recovered in archaeology.[3] As discussed at pp 15–16, Greek contact probably lessened as Carthage sought to control trade past Gibraltar, although the
name 'Tartessos' continued to be attached to the area: see the references there
to weasels and morays. On any view, archaeology shows continued prosperity.

1 López Ruiz (2016) 266–80, more detailed than Celestino and López Ruiz (2016) 111–21.
2 As reported in these later authors, it was where gold and silver came from.
3 We might also think of a port at or near Huelva, at the mouth of the Odel, about 50 miles, a further
day's sail, from the Guadalquivir; again archaeology has not found it.

The Carthaginian settlement of Gadeira (Gades in Latin), Cadiz, is about 30 miles by sea south-east of the mouth of the Guadalquivir. The coastline at Cadiz was quite different in ancient times, and modern maps are not helpful. It consisted of two islands off the mainland. The smaller and more northerly, around the Castilla de San Sebastián and La Vina, became known as to Greeks as Erytheia (see below), and was the port (Strabo 3.5.4; Pliny *NH* 4.120).[4] It may be that Gadeira was the Phoenician name (*'Gdr*) for this island alone. Adjacent was a long island running south-east for some 10 miles to what is now Playa del Castillo (see *BNP* sv). Archaeology confirms its prosperity from an early date: Aubet Semmler (2001a).

It is against that background that we must assess whether Stesichoros, and Strabo's use of him, reflects geography as opposed to poetic imagination. Stesichoros, c 600–550, came from Himera in Sicily (now Termini Imerese, some 25 miles east of Palermo on the north coast), and lived either there or in southern Italy.[5] The quotation is from the *Geryoneis*, his version of the myth of Geryon(eus). A three-headed monster, in the traditional version (Hesiod *Th* 285–94, 979–83) he lived in the misty island of Erytheia in the far west, beyond the Ocean. Heracles' 10th labour was to kill him and steal his cattle.[6] The cited lines show that Stesichoros relocated Geryon's birthplace and residence to the real world. Given his dates, it is likely that he learned about Tartessos from Greek merchants and mariners putting in at Himera, or wherever he was living at the time, who talked about where they were going to trade for silver. Factually, they were going to deal at a port or trading post at the mouth of a river, for silver, and probably tin, which was mined inland and brought down the river. As they described their business, they could well have used 'Tartessos' in a general sense, the place they were going to or coming from. 'We trade at Tartessos' could mean the realm, the port, or the river; at least Stesichoros may have so understood it. At all events, their talk was most likely the reason why he was able to place Geryon in the real world as he did. The 'cavern in a cliff' might be his imagination, but it could have been inspired by reports of the Cave of St Michael at Gibraltar, another milestone on the merchants' route. Gadeira is not Tartessos, and there is no river there. But it was another convenient stop for traders *en route* to or from Tartessos.

While Strabo treated Stesichoros as factual, 'Tartessos' for the Guadalquivir is known only from this poem, and Aristotle, *Meteor* 360b 2, who would have no

---

4 As noted below, Erytheia was originally a mythical island in the distant west.
5 See his biographies in *BNP* and *OCD*.
6 Hesiod, *Theog* 293, names Geryon's cowherd as Eurytion, and the Greek here could be translated as 'Geryon's cowherd'. But it was Geryon, not his servant, who was born in the cliff. A distant island beyond the Ocean was also the location of the three Hesperides, the daughters of Nyx (Night), who guarded the golden apples; one of the daughters was called Erytheia. It was Heracles' 11th (12th in some traditions) labour to steal the apples.

reason to disbelieve his source, Stesichoros or a later author who copied him.[7] Its local name was the Baetis, though that name is first recorded only after the area became Roman: often in Livy (28.2.15, etc), Mela (3.5), and Pliny (*NH* 3.7–11). We cannot recover whether it was the Greek traders who first called the island at Cadiz 'Erytheia', or Stesichoros imaginatively moving the island of myth into the real world; having given Geryon a real birthplace, he needed to give him a real residence.[8]

Understanding the remainder of this text is complicated by two things. One is the lack of information as to the history of the area. At some stage, the area was no longer called Tartessos; later references to Tartessos are to the 'olden days', e.g. Paus 6.19.4. Strabo calls the area Turdetania, occupied by two tribes, the Turdetani and the Turduli; the precise relationship between those tribes was uncertain even in his day.[9] But archaeology does not suggest a violent transition, and it is probable that the Greek merchants were dealing with their ancestors. If Herodotos' *basileus*, pp 14–15, was indeed a hereditary chieftain, his line may have died out. On any view, the Romans called the area Turdetania, after one of the tribes living there. The second stems from the way Strabo uses his sources; he here says 'some say X, Eratosthenes says Y, and Artemidoros more correctly says Z'. It is clear that Pytheas said something about the area, and Eratosthenes used it; it was therefore suspect in Strabo's eyes. But when Strabo says 'some say . . .', can we infer that Eratosthenes had these details (or some of them) from Pytheas, and Strabo had to accept them because others said the same thing?

His source(s) for the river mouth were correct. It is now silted up, but there was once a delta;[10] the Greek merchants who went there reported it correctly. 'Tartessos' as the name for the city, in practice perhaps no more than the trading post where Greek merchants did their deals, could reflect their so calling it, as

---

7 Aristotle treats the Pyrenees as the source of this river flowing south and the Istros, the Danube, flowing east. The river is also called 'Tartessos' in Avienus' archaising poem (Appendix 1) at v 282, and in Steph Byz sv Ταρτησσός, *Tartessos*: 'city in Iberia, after the river, which flows from the Silver Mountain'.

8 That touches on the wider point, that Greeks saw what poets said as factually correct. Erytheia is virtually the Greek word for 'red' (ἐρυθρός, *erythros*). Pliny, *NH* 4.120, offers a rationalisation: the ancestors of the Carthaginians, thought to have founded Cadiz, were also thought to have come from the Red Sea. Herodotos wrongly but understandably imagined that Gadeira was on the mainland, with Erytheia just offshore, in his account of the Geryon myth (4.8.2). Mela 3.47, just before F12, locates Erythia in Lusitania (Portugal), adding that some say it is the home of Geryon.

9 Strabo 3.1.6–7, 3.2.1; generally *BNP* sv Turdetani. Strabo describes Turdetania in some detail; at 3.2.7 he notes the many fish coming from its coast, including morays and tuna (cf Appendix 4 §5); it is clear that he is thinking of their export to Rome (3.2.6). Strabo approved of the Turdetanians, as they had adopted Roman ways: 3.2.15.

10 Modern marine geoarchaeology shows this: Belén Deamos (2009) 197 n 33. Pliny noted its then shore line: 'aestuaria (estuaries) Baetis', *NH* 3.11; Avienus (n 7) v 280 refers to the otherwise unknown Lacus Lagustinus (though Ligystinē is noted in Stephanus of Byzantium as a *polis* near Tartessos).

suggested above. As to the land called 'Tartessis' (feminine), we find the feminine form in Strabo to denote part of a territory.[11] Taking the two references to Tartessis together, it would seem the eastern part of Tartessos up to Calpē, Gibraltar, was thought of as separate in some sense, occupied by the Turduli, with the Turdetani living further west and the Guadalquivir as a natural border. Strabo cites Polybios, who had sailed that coast, as saying that the Turduli lived north of the Turdetani.[12] That could also be true: we do not know if their territory came down to the coast west of Gibraltar; Cadiz itself could have been part of Turdetanian territory. For Strabo, another tribe, the Bastetani, lay further east, beginning at Calpē, Gibraltar.

It would be natural for Pytheas to have noted that he had passed from the Inner Sea, the Mediterranean, into the Ocean, but we have no evidence that he did so, or in what terms. The narrows at Gibraltar, and the names of the two sides, must have been known to mariners from an early date, though it is unclear when Heracles became associated with them.[13] The text is ambiguous as to whether it was Pytheas who first recorded its name, Calpē, or Tartessis as the area around Cadiz.[14] The prosperity of Erytheia, the Greek name for the port of Gadeira, Cadiz, is confirmed archaeologically, as noted above; Eratosthenes' adjective is εὐδαίμων, *eudaimōn* (εὐ, *eu*, 'well', and δαίμων, *daimōn*, a minor god). Roseman (1994) 60 translates 'fortunate' (so Bianchetti (1998) 85, 'fortunata'); the connotation is 'prosperous' or 'wealthy' (as Euboia at Hdt 5.31.2).[15]

For the geographer Artemidoros, see p xix; Strabo naturally preferred him for Spain because, unlike Eratosthenes, he did not rely on Pytheas. Strabo did not realise (or ignored) the fact that Eratosthenes had limited information about the

---

11 Strabo 13.1.7, where Homer's Lyrnessos in the Troad is split into two parts, one called 'Lyrnessis'; 16.4.8, Tēnessis, an area in Egypt. At 14.3.4 Telmessis is the promontory next to the town of Telmessos. For the -essis ending for the female inhabitant, see Herodian *Pros Cath* 3.1.85, e.g. 'Molossis', female Molossian; 'Tartessis' for a female inhabitant (our only other occurrence of the form), Steph Byz sv Tartessos.

12 Polyb 34.9.2 = Strabo 3.1.6.

13 Calpē was the rock of Gibraltar; on the African side it was the mountain Abilix or Abilē, Strabo 3.5.5, 17.3.6, now Jebel Musa. Our earliest accounts of Heracles' labours (see above) do not mention the Pillars. Their earliest known mention is in Dicaiarchos (Polybios 34.9.4, within Strabo 3.5.5), but the Strabo passage shows that there were different traditions about just where the Pillars were: generally Roller (2018) 167–8. There was also a nearby town of Calpē, but Strabo here refers to the rock. Pytheas would certainly know of Calpē.

14 Bianchetti (1998) 117 suggests that 'Tartessis' could be from Pytheas, though her reasoning is questionable: 'una derivazione piteana si può intravedere', 'a derivation from Pytheas can be glimpsed' in Eratosthenes' figures for 7,000 stades from Marseille to Gibraltar and 6,000 stades from the Pyrenees to Gibraltar, Strabo 2.4.4. But whether those could be Pytheas' distances is doubtful: see the Coda p 171.

15 Jones' Loeb has 'Blest', perhaps recalling the myth of the Blessed Isles in the distant west. But both Pytheas and Eratosthenes were practical men; further the Isles of the Blest always have a different adjective, μάκαρ, *makar*, 'blessed', particularly of gods as opposed to men: so Hes *Op* 171, Anon *PMG* 584: generally Gantz (1996) 132–5, which also covers the Elysian Fields.

area: mariners' reports and Pytheas; Artemidoros could access details unknown in Eratosthenes' time. As elsewhere, e.g. F26, he does not specify Pytheas' many lies; and, as with some of his other attacks, e.g. F25 and F31, he gets muddled. It is hard to see why Artemidoros disagreed about Calpē, especially as Strabo accepts it elsewhere about a dozen times, or Tartessis, unless he had a different description of Turdetania which Strabo did not adopt. Pytheas was not wrong for his five days from Cadiz to the Sacred Cape. The name of several promontories, this is Cape St Vincent, near Sagres in south-west Portugal. As the crow flies, it is a little over 160 miles from Cadiz. A sailor keeping to the coast would cover about 210 miles. Strabo's 1,700 stades, 195 miles, however estimated pp xvii–xviii, is reasonable, and could no doubt be done in two days and nights with a favourable wind, by those familiar with the coast, or by the larger vessels with many rowers to supplement wind power in use by the Romans in Artemidoros' time. Pytheas would only sail by day, and had to negotiate an unfamiliar coast: north-west from Cadiz, south-west from Huelva, and north-west from Faro; his timing would depend on the winds. See also the Coda, pp 171–2, for this stretch. Artemidoros was not comparing like with like.

The tides reference is interesting. It seems clear that what Pytheas actually wrote has got mangled (or perhaps deliberately misunderstood). With the aid of Google satellite view, we can see what Artemidoros and Strabo could not know: there are three bays at the cape, and the whole coast for some distance both north and east is a rocky cliff face. The waves would beat against the rocks without much, if any, noticeable diminution in strength, unlike a beach with normal tides. Tides at the cape are also affected because it separates waves flowing north on to south-facing coasts, and east on to west-facing ones. What Pytheas actually noted was how waves behave against a long stretch of sheer rock. Ironically, Strabo was aware that this was so; but he seems to have imagined that the Sacred Cape was a gentle projection onto a beach.[16] Also, F1 shows that Pytheas took careful note of tides in the Atlantic; Artemidoros and Strabo ignore this. It is probable that it was Pytheas who first recorded the cape as 'sacred'. Either he learnt its name from the locals, or he deduced it because there was a shrine or temple to a local deity on the top. If he described the west end of his five-day sail in some other way, we must infer that Eratosthenes assumed that he meant what by his time was known as the Sacred Cape.

The 'easier access' reference throws light both on Pytheas' rate of progress, and how he thought of northern Europe (map 1). The 'access' must refer to routes across the foothills of the Pyrenees linking the Mediterranean to the

16 See 3.3.3: he notes that Aristotle (F680) had pointed out that a rocky coast effectively cancels tides ('receives them roughly and gives them back with equal violence' (trans Jones, adapted)); he disagreed because Posidonios, who had seen the coasts in question, southern Spain and Morocco, had said that they were mostly flat and sandy.

Atlantic, though, as noted on pp 9–10, before he set off we cannot be sure how much he knew about the Atlantic end of the Pyrenees or those routes. On any view, he knew the position when he reached what Strabo calls the northern parts of Iberia, presumably the San Sebastián–Irún area. As proposed in the Coda, pp 171–2, it will have taken him some 5 to 6 weeks in our terms, several phases of the moon for him;[17] his observation underlines how little any part of the north was properly understood in his time. We have no evidence that he expressed the length of this part of his voyage, whether as days or as a distance in stades; but his comment suggests that he tried to make good progress, otherwise the comparison was pointless. Strabo's criticisms are not affected by his (and probably Artemidoros') belief as to the shape of Iberia, including that the Pyrenees ran from south to north: see Appendix 7. His attack on Pytheas here is comparatively mild: his word is ἀλαζονεία, *alazoneia*, literally 'false pretences'.

F35 Strabo 3.4.4

. . . καίτοι ἐμοί γε δοκεῖ δυνατὸν εἶναι καὶ συνηγορῆσαι πολλοῖς τῶν λεχθέντων καὶ εἰς ἐπανόρθωσιν ἄγειν καὶ μάλιστα εἰς ταῦτα ὅσα Πυθέας παρεκρούσατο τοὺς πιστεύσαντας αὐτῷ κατὰ ἄγνοιαν τῶν τε ἑσπερίων τόπων καὶ τῶν προσβόρρων τῶν παρὰ τὸν Ὠκεανόν. ἀλλὰ ταῦτα μὲν ἐάσθω, λόγον ἔχοντα ἴδιον καὶ μακρόν.

But to me at least, it seems possible both to defend much of what has been said, and to revise them; in particular in relation to matters where Pytheas has led astray those who trusted him, in ignorance of regions in the west and the north along the Ocean. But let this drop, as it is a long specialist topic.

The context here is still within Strabo's description of Spain (cf F34). He is arguing that Homer's description of places beyond Gibraltar, with some revisions in the light of later knowledge (or at least beliefs) was correct, so that later geographers were wrong to criticise it. Strabo strongly believed that Homer's geography, including the Ocean that encircled the world, was basically correct: cf p 10. His words here ('west and north along the Ocean') are wide ranging, as if he is repeating his attack on Pytheas for northern Europe in F25 and F26. Of 'those who trusted him', we can identify Eratosthenes and Hipparchos (F24 to F28, F31, F34); Dicaiarchos, Polybios, and Artemidoros did not, to judge from F30 and F34. It was perhaps his antipathy to Pytheas that led him to say that northern Europe east of the Elbe was unknown: see at p 83 and the Note to Appendix 4 §3, though cf the Coda p 183.

---

17 A route along the foothills of the Pyrenees would be over 300 miles, 480 km, on both their north and south sides, and require at least 15 days.

F36 Strabo 4.2.1 = Polyb 34.10.6–7

... ὁ δὲ Λείγηρ μεταξὺ Πικτόνων τε καὶ Ναμνιτῶν ἐκβάλλει. πρότερον δὲ Κορβιλὼν ὑπῆρχεν ἐμπόριον ἐπὶ τούτῳ τῷ ποταμῷ, περὶ ἧς εἴρηκε Πολύβιος, μνησθεὶς τῶν ὑπὸ Πυθέου μυθολογηθέντων, ὅτι Μασσαλιωτῶν μὲν τῶν συμμιξάντων Σκιπίωνι οὐδεὶς εἶχε λέγειν οὐδὲν μνήμης ἄξιον, ἐρωτηθεὶς ὑπὸ τοῦ Σκιπίωνος περὶ τῆς Βρεττανικῆς, οὐδὲ τῶν ἐκ Νάρβωνος οὐδὲ τῶν ἐκ Κορβίλῶνος, αἵπερ ἦσαν ἄρισται πόλεις τῶν ταύτῃ. Πυθέας δ' ἐθάρρησε τοσαῦτα ψεύσασθαι.

The Loire flows out between the Pictones and the Namnitae. Formerly there was a trading station, Corbilo, on this river, about which Polybios, with the fabulous stories of Pytheas in mind, said that those in Marseille with whom Scipio spoke could tell him nothing of value when questioned about Britain, nor the people of Narbo nor those of Corbilo, although these are most important towns in the area. But Pytheas boldly told lies about this.

Strabo's book 4 deals with Keltikē, France; at this point, he is dealing with the area between Nantes and Saint-Nazaire. He now cites from the geography section of Polybios' *History*, as in F30. The Pictones lived in a large area on the south bank of the Loire (Caes *BG* 7.4.6; Plin *NH* 4.108, 17.47); their name survives as Poitiers and Poitou. The Veneti and Namnetes (Caes *BG* 3.9.10; Pliny *NH* 4.107) were on the north bank: cf on F12 n 3, hence Nantes.

The last sentence is probably Strabo's rather than Polybios' comment, but the context as a whole shows that Pytheas put in at Corbilo, a trading station on the Loire estuary, and probably then learnt about the tin trade with an island four days' sail distant, and also its name, *Prettania* (F4, F5); Coda p 172. Strabo noted the Loire as an established point for trade with Britain, within his description of Britain (4.5.2: see on F38). It was almost certainly an important staging post for tin from Cornwall, as noted on F5. Caesar, *BG* 3.8.1, notes that the Veneti had many ships for trading with Britain; the word ἄριστος, *aristos*, 'best' (here translated 'most important'), whether Polybios' word or Strabo's paraphrase, underlines its importance. Although here described conventionally as a *polis*, town or city, it may in fact have been little more than a trading station with a harbour and a few buildings. This would be consistent with its location being unknown, though it is generally accepted to have been on the Loire estuary somewhere to the west of Nantes. This text is our only reference to it; it was not named by Caesar, and his defeat of the Veneti in 56 meant that the trade moved to the Roman settlement at Portus Namnetum, Nantes.[1]

---

1 Caesar's defeat of the Veneti in 56, which included enslaving the population (*BG* 3.14–16) would put an end to Corbilo as a trading centre; Roseman (1994) 66 helpfully notes Bowen (1972) 60 for this; OCD sv Corbilo suggests trading then moved to Portus Namnetum; Bianchetti (1998) 123–4. There is a useful discussion of Corbilo at fr.wikipedia.org/wiki/Corbilo.

Polybios was hostile to Pytheas, as discussed on F30 at p 96 and §2. Narbo (Narbonne) was a long-established trading centre; Hecataios of Miletos (c 500) included the 'Narbaeoi' in his *Geography* (*FGrH* 1 F54), and Strabo was to stress its importance as a major trading emporium at 4.1.6; see Roseman (1994) 66.² It became a Roman colony in 118, around the time of Polybios' death. Polybios will have put in at both Marseille and Narbo on one of his voyages with Cornelius Scipio Africanus to Spain and Africa in the middle of the second century, for which see on F30. Scipio's interest in Britain is not stated, but given his sailing some distance down Africa and up Portugal, it was presumably trade related; for Britain it would particularly be tin. But traders were always reluctant to say more about their sources than they had to: Roseman (1994) 66 notes that Gallic traders would not tell Caesar much about Britain, *BG* 4.20–1, and cf the stories Herodotos recorded, p 8. So Celtic traders would want to say little or nothing about Britain or their trade with it, at any rate to a Roman official such as Scipio. He, and Polybios, could only know of Corbilo by hearsay: we may assume that there happened to be one or two people from Corbilo in Narbonne or Marseille when Scipio was there, as tight-lipped as the others (Walbank (1979) 612; Roseman (1994) 66). It was then easy for Polybius to turn his one or two Corbilese informants into a general statement that Britain was unknown in Corbilo, and therefore Pytheas lied about it.³

F37 Strabo 4.4.1

Ὀσίσμιοι δ' εἰσίν, οὓς Ὠστιμίους ὀνομάζει Πυθέας, ἐπί τινος προπεπτωκυίας ἱκανῶς ἄκρας εἰς τὸν ὠκεανὸν οἰκοῦντες, οὐκ ἐπὶ τοσοῦτον δὲ ἐφ' ὅσον ἐκεῖνος φησι καὶ οἱ πιστεύσαντες ἐκείνῳ.

. . . *[and] there are the Osismioi, whom Pytheas calls the Ostimioi, who live on a headland that projects quite far into the Ocean, though not as far as he and those who have trusted him say.*

Like F36 and F37, this is from Strabo's geography of France. He is here covering Brittany; he has just noted the Veneti, who lived north-west of the Pictones and Namnitae of F36, in the Morbihan area around Vannes, and now mentions the Breton tribe whom Pytheas recorded. Eratosthenes had copied Pytheas both for the name and the headland: see on F26; the name of the headland is there given as 'Cabaion'. Strabo describes the cape as projecting ἱκανῶς, *hikanōs*, into the Ocean. *Hikanōs* means 'sufficiently', hence the translation here, 'quite far'. This is vague, and while he states his usual disbelief of Pytheas, he does not go on to

---

2 Cf portuslimen.eu/site/narbo-martius/.
3 For the extent of Polybios' sailing in the Atlantic, see on F30 with n 1. This text is not evidence that he got anywhere near the Loire.

say what Pytheas said, or how others described it more correctly.[1] The problem of fitting in Brittany to his perception of the coast of Keltikē is discussed in Appendix 7. 'Those who have trusted him' will include Eratosthenes; as we see in F26, Strabo disagreed with how Eratosthenes worked out his dimensions for the length of the world, including at its western end. The tribal name as printed here is a possible emendation of Strabo's MSS; as noted also on F23, F25, and F26, its spelling has got mangled in transmission. The problem of recovering how Pytheas actually spelled it is discussed in Appendix 5.

F38 Strabo 4.5.5

Περὶ δὲ τῆς Θούλης ἔτι μᾶλλον ἀσαφὴς ἡ ἱστορία διὰ τὸν ἐκτοπισμόν· ταύτην γὰρ τῶν ὀνομαζομένων ἀρκτικωτάτην τιθέασιν. ἃ δ' εἴρηκε Πυθέας περί τε ταύτης καὶ τῶν ἄλλων τῶν ταύτῃ τόπων ὅτι μὲν πέπλασται, φανερὸν ἐκ τῶν γνωριζομένων χωρίων· κατέψευσται γὰρ αὐτῶν τὰ πλεῖστα, ὥσπερ καὶ πρότερον εἴρηται, ὥστε δῆλός ἐστιν ἐψευσμένος μᾶλλον περὶ τῶν ἐκτετοπισμένων. πρὸς μέντοι τὰ οὐράνια καὶ τὴν μαθηματικὴν θεωρίαν ἱκανῶς ἂν δόξει κεχρῆσθαι τοῖς πράγμασι . . . τοῖς τῇ κατεψυγμένῃ ζώνῃ πλησιάζουσι . . . τὸ τῶν καρπῶν εἶναι τῶν ἡμέρων καὶ ζώων τῶν μὲν ἀφορίαν παντελῆ, τῶν δὲ σπάνιν, κέγχρῳ δὲ καὶ ἀγρίοις λαχάνοις καὶ καρποῖς καὶ ῥίζαις τρέφεσθαι· παρ' οἷς δὲ σῖτος καὶ μέλι γίγνεται, καὶ τὸ πόμα ἐντεῦθεν ἔχειν· τὸν δὲ σῖτον, ἐπειδὴ τοὺς ἡλίους οὐκ ἔχουσι καθαρούς, ἐν οἴκοις μεγάλοις κόπτουσι, συγκομισθέν των δεῦρο τῶν σταχύων· αἱ γὰρ ἅλως ἄχρηστοι γίνονται διὰ τὸ ἀνήλιον καὶ τοὺς ὄμβρους.

Concerning Thule our information is even more uncertain because of its distant position; for they place it as the most northerly of named [places]. But what Pytheas has said about it and other places in that area is fabricated, as is clear from places that are known about. For he has told lies about these places for the most part, as has already been said, so he is clearly lying even more about remote places. However, with regard to celestial matters and mathematical theory, he might seem to have used facts adequately . . . for those living near to the cold zone . . . some cultivated fruits and livestock are completely lacking, and others scarce; they live on millet and wild herbs and fruits and roots. They have grain and honey, and they make a drink from them. As to grain, since they do not have clear sunshine, they winnow it in large buildings, having brought their ears of grain there; for threshing floors are useless because of the lack of sunshine and the rain.

Like F36 and F37, this comes from book 4, Strabo's geography of France: he includes a short section on Britain and Ireland, 4.5.1–5, and concludes with the

---

1 Nothing is to be gained by noting the linguistic detail that Strabo's, and perhaps Eratosthenes', word at F26 is ἀκρωτήριον, *akrōtērion*, there translated 'cape'; here, perhaps from a different source (Artemidoros?), the word is ἄκρα, *akra*, 'headland'. There is no practical difference.

Alpine region, broadly from Switzerland to Monaco and Genoa. He then turns to Italy in book 5. He says that Britain is triangular, opposite Keltikē (4.5.1: see Appendix 7). There are four main sea-routes to it.[1] Most of it is flat, with many hilly districts. His sources were reflecting the south, as known from trading and Caesar's abortive invasions; it shows no knowledge of mountains such as Wales or the Pennines. Their way of life is partly like the Celts', but more barbaric. He did not understand that Britons were also Celts (cf on F28). They live in huts in forest clearings. They grow grain and have cattle; they have milk but make no cheese, and do not engage in cultivation or agriculture.[2] Like others (see on F4 and F12), he notes their chariots (4.5.2). Unlike elsewhere in his *Geography*, he offers no place names. After claiming that Caesar's invasions had effectively pro-cured the whole island for Rome by the submission of their chieftains (4.5.3), he says that there are small islands around it, which he does not name (and so he did not use Pytheas there), and Ireland, for which see on F31.

It is unfortunate that there is a *lacuna* in our MSS of Strabo. His ambiguity as to whether Thule exists or not is noted on F25. At F31 he said 'ἱστορῶ', '*historō*', 'I can record' nothing from others; here he says that his ἱστορία, *historia*, 'infor-mation' is uncertain, implying that it exists. *Historia* basically meant 'enquiry'; it came to mean information from enquiries or a written account of that (hence 'history' in our sense).[3] Here, he states definitely that Pytheas' account of it is lies; it does not exist. He then acknowledges Pytheas' science, as he does in F39. Mette (1952) 18–19, 36–7, proposes that it was Posidonios' assessment of Pytheas which Strabo adopted. Posidonios' works included astronomy: see on F2 with n 2; in the introduction to his *Geography*, 1.1.13, 20–1, Strabo says how important astronomy and geometry are to geography. We may suspect that some Pytheas astronomy has dropped out of the text; possibly, given the context, the summer tropic–arctic circle point he had mentioned in F31.

I have indicated a further gap after 'for those living near the cold zone'. Whether or not we have lost one of Pytheas' astronomical details, either before or after this phrase we have lost words to the effect that Pytheas refers to those living in the north. It is often said that the diet is that on Thule, and the passage is used to locate it, particularly noting that bees cannot live further north than the latitude of the Shetlands or Bergen: e.g. Roller (2018) 202, (2006) 86; Roseman (1994)

---

1 The usual passages to Britain were from the Rhenus (Rhine), the Sequana (Seine, Le Havre), the Liger (Loire), and the Garumna (Garonne, Bordeaux); to unnamed British ports. It does not mean that there was no cross-channel traffic with the Morlaix area, but, as noted on F5, it would be cheaper to use the Loire estuary.
2 The words are κηπαῖος, *kēpaios* (adj), relating to gardens or cultivation, and γεωργικός, *geōrgikos*, relating to agriculture.
3 Strabo basically refers to written texts, but as noted at p 1, Virgil could only refer to Thule if it was spoken about, and *historia* certainly included oral input. Our 'history' goes back to Herodotos' use of the word in the proem to his book.

134–8; Bianchetti (1998) 170–3. But making all due allowance for the *lacuna*, this is not what Strabo says. He concedes that Pytheas could have been right in describing the diet. Therefore it must be the diet of real people living in a real place, and in practice that must mean Britain. Roseman (1994) 135 argues that because Strabo uses the perfect participle of καταψύχω, *katapsychō*, to cool or to chill, to describe the zone, he is referring to an area subject to continual cold but not necessarily freezing.[4] Her use of 'freezing' introduces a detail that is not in Strabo; he has the simple concept that the far north is uninhabitable because of the cold (ψῦχος, *psychos*, e.g. 2.1.13, 2.5.6, F31, etc); he accepts that people can live in the northern parts of the habitable zone. Pytheas did not speak of habitable or uninhabitable zones. Strabo is reluctantly accepting that Pytheas is right about the diet, at least in some part(s) of Britain. Strabo placed Britain south of Ireland and therefore within the habitable zone (maps 3a and 3b, p 161).

F30 §2, however we understand it, shows that Pytheas spent some time on land, though it is not clear whether the diet is from just one community or based on what he learnt in several places. Diodoros was able to access details about the way of life in Britain from source(s) other than Pytheas: see the passage printed at the end of F4; the reference to grain being stored in roofed buildings might perhaps go back to him.

It is helpful to compare Pytheas' description of the British diet, and what Strabo says as noted above, with modern archaeology; the brief summary here is based on the Iron Age Diet section in scarf.scot (search 'Iron Age Diet') and Jay and Richards (2007). It is not the diet in the Northern Isles, as neither mention fish, regularly recovered on Scottish islands but rarely in England. It is impracticable to comment on 'fruits', cultivated or wild; both are likely. Neither mention nuts. As to livestock, their bones are so commonly recovered that we can only say that in some areas sheep, goats and pigs might be more common than cattle. Strabo's comment about not making cheese (whether from sheep's milk or cows') is surprising, and probably wrong; certainly his statement about cultivation and agriculture is wrong; vegetables were certainly grown: Jones (1996) 33; cf below.[5]

Pytheas' mention of millet is problematic. It was not common in Greece but, as pointed out at Dalby (1996) 90, Herodotos (1.193.5, 3.100, etc) and Xenophon (*Anab* 1.2.22) mention it in contexts where it would be understood by their

---

readership.[6] But it is almost certain that it was not grown in Britain.[7] On the other hand, oats were also rare in Greece. Pytheas might know the word for it (βρόμος, *bromos*), but not what they looked like. They were grown in Iron Age Britain, but not widely.[8] It is suggested that he meant oats (e.g. Cary and Warmington (1929) 39; Thomson (1948) 145; it is possible that he visited a community which did grow oats. Against that, he goes on to note grain, which suggests that his 'millet' was not a grain crop. It is attractive to note Roseman's suggestion, (1994) 136, that he saw wild plants whose seeds looked like millet seeds.[9]

The next item is also problematic. Pytheas has λαχάνος, *lachanos* (here in the dative plural), which means both herbs and vegetables, plus an adjective. Most MSS have ἄλλοις, *allois*, 'other', but one MS has ἄγριος, *agriois*, 'wild', which some editors prefer (most recently Radt (2002) 526), adopted here; the context requires it. Whatever Pytheas meant by 'millet', it was not an herb or vegetable. As noted above, vegetables were grown, but perhaps not to any extent where he went. The drink could be either mead or beer. Both are concoctions of grain and honey; Pytheas used the neutral πόμα, *poma*, drink.[10] But for Greeks, who drank wine, mead was a medicine; if it was the Britons' social drink, we would expect Pytheas to have commented.[11] Celts brewed beer, as noted below; Pytheas saying 'drink' is probably because his Marseille dialect, at any rate in his time, did not have a word for beer. Herodotos had reported that the Egyptians had a drink 'made from barley' (2.77.4), but without offering a name. Our earliest text for 'beer' is Theophrastos, about a generation younger than Pytheas. He said that it was an Egyptian drink made from barley and wheat, and their word for it was ζῦθος, *zythos*. This became the Greek word for beer; but we cannot assume that

6 The Herodotos and Xenophon references in Dalby at 90 are wrong.
7 The cereal diet of Iron Age people can be deduced from the residual protein in their bones. In photosynthesis, plants typically use one of two metabolic pathways, known as C3 and C4; they differentially incorporate the stable isotopes of carbon, ^{13}C and ^{12}C, as CO_2. The distinctive $^{13}C{:}^{12}C$ ratios persist after the plants have been eaten and the carbon incorporated into human tissues, and so provide evidence of diet. Wheat and barley are C3 plants; millet is C4, which is found in Iron Age bones in Europe, but not Britain: Jay and Richards (2007) 172. It is not mentioned in Jones' lists of cereal crops, (1996) 31–3.
8 'Oats (*Avena spp.*) have a low but persistent presence in Iron Age assemblages but often these cannot be determined to species and many may be simply weeds of the barley crop. At some sites cultivated oats (*A. sativa/strigosa*) have been positively identified' (scarf.scot, §4.2). In the several sites in Jay and Richards (2007), there were wheat and barley but no oats.
9 'Millet seeds are very small, grow in clusters along a stem, and are spherical. . . . it would seem more reasonable that he was referring instead to weed seeds used as food'.
10 Briefly, mead is fermented honey, and any added grain would be for flavour (and could be made without boiling the honey liquor). Beer is fermented grain (normally barley), and needs boiling; the honey would be added for extra sweetness.
11 The word, ὑδρόμελι, *hydromeli*, only occurs in medicinal writers (with the theoretical exception of Athenaios 2.61d, citing the medical writer Diphilos). But hydromel has been suggested: see Bianchetti (1998) 171.

it circulated in Pytheas' time.[12] Note that Pytheas uses σῖτος, *sitos*, for the grain, the general word for both wheat and barley. Britons' Celtic cousins in Gaul are reported as using both for their beer. In the context of their preferring wine over their traditional beer, Posidonios F67 E–K says that it was made from wheat, and Diodoros 5.26.2 from barley; in each case the specific word is used.[13] We naturally think that beer is made from barley, but Posidonios was a scientist, and if he is accurately reported he would mean corn.[14] Perhaps Pytheas found both grains being used, whichever was more abundant in different places.

When he refers to threshing, his word, στάχυς, *stachys*, normally refers to ears of corn in surviving texts, but it could refer to the ears of any grain: Chantraine (1968–1980) 1045; Frisk (1972) 779. It is interesting that he noted the rain; British weather seems to have become notorious, as it was one of the few things about Britain Strabo could include in his British section (4.5.2). The threshing of grain here, and its storage in Diodoros' source for the passage printed at the end of F4, rather contradicts Caesar's statement, *BG* 5.14, that Britons were mainly pastoral, as noted by Roseman (1994) 137. But Caesar's knowledge was limited to the south-east of the island.

This reference to Pytheas' astronomy can be read as evidence that Strabo had his actual book: what was his source for the diet in Britain? His usual references to Pytheas are second-hand from Eratosthenes and Hipparchos, where he complains that they (mis)used him for astronomy and geography. It is not a topic that seems to fit easily into how we believe either of them wrote their geographies. Except possibly for the phrase 'their way of living is modest', the passage in Diodoros (see above) does not touch on diet; but Strabo could access sources which did, and he reported as just noted; in addition he had another source, with other details, which was specifically from Pytheas. The tenor of his references to Pytheas elsewhere is that he read him as cited by others, and on balance he also got Pytheas' British diet from an intermediate source.

F39 Strabo 7.3.1

... διὰ δὲ τὴν ἄγνοιαν τῶν τόπων τούτων οἱ τὰ Ῥιπαῖα Ὄρη καὶ τοὺς Ὑπερβορείους μυθοποιοῦντες λόγου ἠξίωνται, καὶ ἃ Πυθέας ὁ Μασσαλιώτης κατεψεύσατο τῆς παρωκεανίτιδος, προσχήματι χρώμενος τῇ περὶ τὰ οὐράνια καὶ τὰ μαθηματικὰ ἱστορίᾳ. ἐκεῖνοι μὲν οὖν ἐάσθωσαν.

---

12 Philologists doubt that it is an Egyptian word, because no comparable Egyptian word is known, and it is more likely a Greek formation from from ζύμη, *zymē*, yeast: Chantraine (1966–77) 401, Frisk (1960–72) 616. But that does not mean that the word was known in Marseille in Pytheas' time.
13 Πυρός, *pyros* (corn); κριθή, *krithē* (barley).
14 Our source is Athenaios 4.152c. The accuracy of Athenaios' citations is considered in Pelling (2000); for Posidonios 174–5; generally 188–90.

. . . because of ignorance of these regions, those who make up stories about the Ripean Mountains and the Hyperboreans are believed, as with the lies Pytheas of Marseille told about land on the Ocean, using the excuse of enquiring about astronomy and mathematics. These men should be disregarded.

Strabo has just ended his comparatively short description of Germany, which for him essentially ended at the Elbe, the land beyond being unknown (see Note to Appendix 4, §3). He now moves on to what lay further east, starting with the Getae, who lived in Romania and Bulgaria. He concludes his section on them by saying that he cannot speak accurately of their boundaries, and continues as above. The existence or otherwise of the Ripean mountains and the Hyperboreans is noted at p 11; it is interesting that Strabo rejects them, while Mela accepts them (see on F13), and Pliny regularly refers to both. Strabo notes them only when quoting earlier writers, the Hyperboreans at 1.3.22, 11.6.2, and 15.1.57, and the mountains at 7.3.6. The mention of both here is appropriate; he is about to speak of other imaginary places in the north. So another attack on Pytheas' veracity about the north fits in nicely.

From our perspective, in the context of the Ripeans and Hyperboreans, the 'land on the Ocean' should logically refer to what Pytheas wrote about the Baltic, such as the Metuonis coast of F21. But with that touch of rhetorical illogicality which we have met in Strabo elsewhere, it is better interpreted as a generalised attack on much of Pytheas' account for the north; perhaps not of Britain, but the coast from Calais to the Baltic. As noted above, Strabo disclaimed knowledge of what lay beyond the Elbe. On the one hand, he is saying: I refuse to cite Pytheas for the further reaches of Europe; on the other I accept that he reported some astronomical details for the north.[1] Perhaps his thinking was: Pytheas was somewhere in this unknown territory, and used his astronomy there, but I refuse to believe that the place exists, at least in the way he describes it. He may also have had Pytheas on Thule in mind. As elsewhere, he does not tell us what Pytheas' lies were. The problems of deciding what astronomy Pytheas actually did in the north are discussed in Appendix 2; the Coda, p 176, proposes that he would be in the Denmark area at a summer solstice. Such of his geography for the north as survives is noted in the Coda, pp 181–3.

---

1 We have to say 'the further reaches of Europe', because we have no evidence that Pytheas recorded rivers, either as such or by a name.

APPENDICES
and
CODA

Appendix 1

THE ALLEGED MASSILIOT
PERIPLUS

Around 350 AD, Rufus Avienus, a Roman official of some literary abilities, wrote a long poem called *Ora Maritima* ('Coastal Shores').[1] The first part survives, describing the Atlantic coasts of Spain and Portugal, with names and details not recorded elsewhere in surviving literature; it mentions Britain under the name 'Albion', and Ireland, and then deals with the European coast from west of Gibraltar to Marseille. It has been proposed that its details go back to an otherwise unknown Massiliot *Periplus* (for the word, see p 7 n 1) of the early sixth century. The topic is too wide for a full discussion here, but it needs consideration for its implications for Pytheas: if it existed, it is argued that parts of northern Europe, including Britain, would already be known in Marseille before Pytheas set out; alternatively, was Pytheas a source for Avienus?

The basic suggestion is that when Greeks such as Massiliots traded with Tartessos, they picked up knowledge of the Atlantic from the Tartessians, who spoke of their own trading links there; and this unknown author duly recorded it.[2] It was then said to have been expanded in Hellenistic times by a second unknown author, and it was this version that Avienus used.[3] It is here proposed that (a) no such *Periplus* existed; (b) whatever Massiliots (or any Greeks) picked up when trading with Tartessos, little, if anything, will have still circulated in Pytheas' time; and (c) if he did hear of any such details, he did not associate them with what he found.

---

1 For his family and proconsulships, see Matthews (1967). The considerable literature on his *Ora Maritima* is in Smolak (1989) 324–5.
2 For Greek trading with Tartessos, see pp 14–16 and on F34. The Marseille connection is assumed because although Herodotos 1.163 says that the Phocaeans were the first to trade with Tartessos, it is inferred her daughter colony Marseille participated. Also a Phocaean layer is inferred because of place names in the poem with the ending -ussa, (Greek -ουσσα, -*oussa*), as opposed to the normal feminine with one *s*, -ousa (Roller (2006) 11; cf Celestino and López Ruiz (2016) 88). While the -oussa form is Ionian, at least in poetry (Buck (1955) 70), there are some 17 places names in -oussa in the western Mediterranean (Dion (1977) 26–30)). At v 195, there are the Cynetes in south-west Spain. This tribe was known to Herodotos as Cynesioi (2.33.3) and Cynetes (4.49.3); they are the furthest known tribe to the west, located near the 'city of Pyrene'.
3 Schulten (1922, 1955) built on an *alte Periplus* proposed by Müllenhoff (1870) 73–203; the alleged *periplus* at 83–203; Stichtenoth (1968); Murphy (1977) v–vi.

DOI: 10.4324/9781003181392-5   123

1 There are three main objections to this alleged Massiliot *Periplus*. Firstly, it presupposes two works, the original and the expanded version, both of which were totally unknown to any other writer: e.g. Herodotos, Hipparchos, Strabo, Pliny. Making all due allowance for how books were published and circulated, this in itself is improbable. Secondly, Avienus' poem contains factual errors. The main ones are:

(i) Avienus says: 'Cadiz, formerly called Tartessus' (v 85). This is wrong: they were distinct: Cadiz is some 30 miles by sea from the mouth of the Guadalquivir. Even if Cadiz was once politically part of or subject to Tartessos (see on F34), they were always separate places. It cannot be an error by the Greek traders; they would know that they were two places, passing Cadiz *en route* to Tartessos.[4]

(ii) He mentions a headland Oestrymnis, overlooking the Oestrymnic bay, containing the Oestrymnides islands, rich in tin and lead (vv 90–101). The similarity to the name of the Breton tribe in F26 suggests that they are in the same area; but he places them next to Gibraltar.[5]

(iii) He also gives this tribe coracles (vv 102–6) because they lack wood; that is nonsense. Whether these lines refer to Brittany, which the context requires, or Britain, both were well wooded. They are said to use these coracles to trade with Ireland, the 'sacred island', two days away. The island of the 'Albiones' is nearby (vv 107–11). Again, unreal: Brest to Cork, for instance, is at least 300 miles and at least three days and nights in a normal boat. Further, 'sacred island' is a play on words: in Greek, Ireland is Ἰέρνη, *Hiernē*, and would sound like ἱερὰ νῆσος, *hiera nēsos*, 'sacred island': see Freeman (2001) 29–30.[6] It is odd that 'Albion' is treated as less important than the more distant Ireland. On any view, the use of this name for Britain must be on metrical grounds; 'Britannia' is metrically impossible.

(iv) The important port of Emporion, founded by Marseille (Empúries, Spain: p 6), is not mentioned in the stretch between Gibraltar and Marseille; any *periplus*, even if not by a Massiliot, must have included it.

(v) A Massiliot author may be ruled out by the description of Marseille in terms appropriate to the lagoon at Martigues, an old Celtic settlement and later the Roman Maritima Avaticorum, and totally wrong for Marseille (vv 561–2): see Berthelot (1934) 15–17 and esp 129–30. The two are some 20 miles apart.

---

4 See on F34 for the possibility that over time, trade was concentrated at Cadiz; later generations might therefore run the two places together.
5 He also has other confusing details with this name: Tartessos used to trade with the islands (vv 107–16), and beyond the islands land that used to be occupied by Ligurians, another puzzle since Ligurians lived between Marseille and Genoa.
6 Avienus has many other sacred islands and other places apart from those noted above.

(vi) There are other difficulties with Avienus' description which can be only briefly noted here.[7]

Making all due allowance for the possibility of Greeks misunderstanding their Tartessian informants, and the later imaginative manipulation of his material by Avienus, the cumulative effect makes it difficult to say that it all goes back to this unknown sixth century author. Points (iv) and (v) can be avoided by arguing for a non-Massiliot author.[8] That does not meet the third objection, that it requires us to assume a *Periplus* at a date when written prose of any sort was still unknown (if early in the century) or unusual (if later).[9] If we ignore that point, it remains unclear why the Mediterranean coast from Marseille westwards needed a written text in an essentially oral world; Massiliot mariners knew it perfectly well; and no one suggests that they ever had plans to take over the Atlantic trade from Tartessos.

More recently, enthusiasm for it has waned. Freeman (2001) 31–3 summarised the arguments for and against and felt unable to decide. But Celestino and López Ruiz (2016) 88–91 are surely right in saying that there is nothing in the poem that could not have been found in texts from Hellenistic times onwards, once Spain, including the Tartessos area, fell under Roman rule. That substantially explains the problems noted above. For the coasts from Huelva and Cadiz to Marseille, vv 497 onwards, by Avienus' time there were a large number of *periploi* and geographies from which he could get details: cf on F9.[10]

2 The *Periplus* may be an ingenious academic construct, but the idea behind it merits consideration. It is entirely reasonable to say that when Greeks traded with Tartessos, they would pick up information there, which would then circulate orally back home. Indeed, it underlines the point noted above, that Pytheas lived in an oral society. They might find local myths more interesting than

---

7 One is the confusion of the Cassiterides, the theoretical tin islands (pp 8–9), with the Oestrymnides, wherever we locate the latter. His Ophioussa (one of the -ussa place names of n 2) is problematic: he effectively identifies it with Iberia, and says that the Oestrymnici used to live there, but were driven out by a snake (vv 152–5). This cannot be a Tartessian story; it is play on words, ὄφις, *ophis* being Greek for snake. Ophioussa was a common name, usually an island: Formentera (Strabo 3.5.1, Ptolemy *Geog* 2.6.73; cf Pliny *NH* 3.38); a settlement on the Dniester near the Black Sea (Strabo 7.3.16, Pliny 4.82; Ptol 3.10.8); one of several earlier names for Rhodes, per Strabo 14.2.7, Pliny 5.132. At Pliny 4.61, it is an island off Crete, and at 4.65 one of the Cyclades, perhaps the same island; at 5.161, an island in the Propontis. Ovid, *Metam* 10.229, with poetic imagination, equates it with Cyprus.
8 E.g. Roller (2006) 9–10 ('early *periploi*', 10); Zehnacker and Silberman (2015) 330 (a 6th-5th century Greek *periplus*); Cunliffe (2017) 282, 284, who proposes a Massiliot author ('sixth to fourth centuries', 282; 'probably . . . sixth century', 284). It was assumed by Rivet and Smith (1979) as noted on F17 n 1.
9 We must distinguish between the ability to read and the writing of books and the keeping of archives: writing was used mainly for epitaphs and laws. Books only started to be written from the mid fifth century in Ionia.
10 See also Gonzalez Poncé (1993), (1995) against a *Periplus*. Note that the Spanish details are not likely to have been learnt by Avienus in person: his second proconsulship was in Africa, not Spain: Matthews (n 1) 489–90.

distant headlands to which they did not intend to go, but for present purposes we can assume that they did learn all about the Atlantic and the places noted above. Would not Pytheas have learnt this before he set off?

The difficulty with that argument is the evanescent nature of oral traditions, what has been called the floating gap or hour-glass effect.[11] Typically, at any given time, both polis and family traditions covered genealogies and foundation myths, but little or nothing more recent until details from the last three generations.[12] What we might infer was spoken of in Marseille in 550 about the distant north as learnt in Tartessos would have largely passed out of memory by 450 and more or less completely by 350.

3 As discussed at pp 15–16, it is hard to know how regular Massiliot contact with Tartessos had become over the two or three generations before Pytheas. But even if we assume that significant contact continued, and Tartessian stories about the north still circulated in Marseille, there are no traces of this in Pytheas, and nothing from which we can infer that he noted them. That point is equally valid if one assumed that an old *Periplus* existed, a copy of it was available in Marseille, and he had read it. It is hard to see that any of the names in Avienus are reflected in what we know Pytheas wrote. While we may assume that mariners spread stories wherever they went, we may note that Herodotos did not know of any of Avienus' details for the little he could say about Iberia.[13] What Pytheas said about the 'Ost-' tribe in Brittany, F26, and the coracles in Britain (see on F19), was from his autopsy. He named Ushant (F27) as such, not as an island of his Ost- tribe. There is no trace of Avienus' 'tin islands', which he associated with Brittany, in Pytheas' factual account of Cornwall, F5. Even if we assume that the name 'Albion' already circulated in Marseille, he did not connect it with the island he called *Prettania* (F5, F18). For his ignorance of Ireland, see the Coda, pp 181–2.

4 There are two final points. Two men did explore the Atlantic before Pytheas, Euthymenes down Africa and the Carthaginian Himilco. But, as noted on pp 12–13, Pytheas could not have learnt about the north from either. Avienus does mention Himilco, with a fantastic story of him being becalmed for four months in what appears to be the Bay of Biscay by sea monsters. The second point is whether Pytheas himself was a source for Avienus, as proposed, e.g. by Carpenter (1973) 199–214 (who assumes that a *Periplus* existed); but only the coracles, with the imaginative addition of there being no timber available, and moving them from Britain to France, is a realistic possibility. As noted above, Pytheas' Ost- tribe is in Brittany, not near Gibraltar; Britain was *Prettania*, not Albion; he did not know the name *Hiernē*, and so could not have spoken about a Sacred Island. Avienus may be a fascinating author, but he is irrelevant so far as Pytheas is concerned.

11 For the expressions see Thomas (2019) 49; Thomas (2001) 198–9.
12 While much of our material relates either to Athens (Thomas (1989), Thomas (1992), esp 108–13, 235–6) or Herodotos (Thomas (2001)), the hourglass shape of memory applied all over the Greek world: Thomas (2019), esp Introduction, 21–8, and chaps 1 and 2.
13 Other than the Cynesioi tribe mentioned in n 2, and the reference to weasels at p 16.

Appendix 2

PYTHEAS' CONTRIBUTIONS
TO ASTRONOMY

Pytheas had a good grounding in astronomy, including terrestrial astronomy, as noted at pp 19 and 22. It is easy to propose that he used this knowledge during his voyage, but it is hard to work out from what survives just what that amounted to in his book. Further, it is worth repeating and stressing a note in the Preface: we must not attribute to him later knowledge, both of data and mathematics. Citations from later authors such as Geminos are misleading if they are read as reflecting what Pytheas knew or wrote nearly 300 years earlier.

1 For the reasons discussed on F8, we should be cautious about treating his mention of the pole star as reflecting his astronomical, as opposed to his mariner's, knowledge. Anything else he said about constellations is lost: see on F2.

2 Strabo refers several times to Hipparchos putting Byzantium on the same latitude as Marseille because his *gnomon* reading for Byzantium was the same as Pytheas' reading for Marseille, taken at the same time of year: see on F25. Pytheas' voyage would have started in spring: cf the Coda pp 171–4, and it is likely that his *gnomon* reading was taken at the spring equinox just before leaving. If Hipparchos compared summer solstice readings, we have to assume that Pytheas took a summer reading in Marseille independently of his voyage, but chose to record it in his book. A *gnomon* reading is simply the length of the sun's shadow. You need only two rods marked into divisions, one of known length held vertically (the *gnomon*), one laid on the ground to measure the sun's shadow. The divisions were presumably the *daktylos*, a finger [width], 0.76″. As noted on pp 19–20, Eudoxos understood that a *gnomon* reading at the summer solstice told him the ratio of day to night on that day: thus a reading of 5:3, where the *gnomon* is 3 units long and the shadow 5 units, meant that 5/8ths of the day is daylight and 3/8ths night, 15 and 9 hours in modern terms (Dicks (1970) 21). With trigonometry, that can be turned into the latitude of the place (Dicks *op cit* 19–23). What Pytheas learnt was probably that any *gnomon* readings at stated times of the year would be of future use. Pytheas lived when astronomy was still finding its way in terms of lengths of daylight and latitude;

DOI: 10.4324/9781003181392-6

he did not have the knowledge, particularly trigonometry, and also more data, that Hipparchos had.

3 Even so, it is easy to infer that he would take other readings as he went; it is a possible inference from the statement in F39 by the usually hostile Strabo, that Pytheas did something, apparently in the Baltic, which he (Strabo) calls 'astronomy and mathematics', coupled with his concession in F38 of Pytheas' competence in those areas. He would need to take only two marked rods as just described. Possible locations for these readings are noted in the Coda, p 180; but they would only be of value to a future astronomer if he also gave their locations; and the nearest we have to that is the distance from Cadiz to Cape St Vincent, F34. However he may well have said that he overwintered in the Orcades, here the Shetlands in our terms (cf on F19); if he did so, and if he took a winter solstice reading, then as noted on F28. Hipparchos could use it as a cross-check against his own calculations.

4 That having been said, it is hard to find evidence for these other readings. Eratosthenes took some geography from Pytheas: certainly for Spain, as we see in F34, and probably for Britain and northern Europe. But the only astronomy he can be shown to have taken from Pytheas was using the latter's statement that the summer tropic and arctic circle coincided at Thule to establish its latitude, as noted on F25. Whatever other astronomy Eratosthenes offered for western and northern Europe is lost. He recorded the length of daylight at places along the meridian through Egypt and on to Byzantium, and probably proposed a latitude through the Dnieper, F31. But none of that depended on Pytheas. We cannot see Pytheas behind such little evidence as we have as to what he offered for Spain, France and Britain.[1]

5 Hipparchos' geography for the north is lost; whatever he said about latitudes there can only be inferred by reference to his daylight figures and winter sun heights; the question for us is whether any are from Pytheas. At 2.5.34 (F39 Dicks) Strabo says that Hipparchos compiled a list or table of latitudes with different celestial phenomena for each. Strabo did not copy this, but noted individual readings in various places in books 1–2; they are our main source for recovering it. Dicks (1960) 154–64 analysed the passages, and inferred that there were in fact two tables, one with astronomical phenomena,[2] and one with the length of the

---

1 The details for Egypt and north to the Dnieper are largely treated as fragments of both Eratosthenes and Hipparchos: Eratosthenes F34, F59, F60 with commentary Roller (2010) 151–4, 169–72; Hipparchos F32, Hipparchos F46–52, 56–60 with commentary Dicks (1960) 172–92. If Eratosthenes extended his Dnieper latitude west through Keltikē, it would not depend on Pytheas. Locating the Roxolani 4,000 stades north of the Dnieper (see on F31) is not from Eratosthenes: Roller (2010) 153.
2 E.g. Strabo 2.5.36, F47 Dicks: at Syene almost the whole of the Great Bear is visible, 'except the legs, the tip of the tail, and one of the stars in the square' (trans Dicks); Strabo 2.5.41–2, F56–57 Dicks, in areas north of Byzantium Cassiopeia is partly or wholly visible. The main Hipparchos fragments are F46–F61 Dicks, (F61 Dicks = our F29), with his commentary at (1960) 172–92.

longest day: see at 162–3.[3] Whether one table or two, we can say that it or they had the lengths of the longest day at the summer solstice, mostly in exact integers, where possible located at a specific place, and the latitude of the place, and in northern locations the height of the sun at the winter solstice in astronomical cubits (one cubit = 2°).[4] Part of his table(s) is set out on F28.

6 Hipparchos repeated and perhaps refined Eratosthenes' figures at the summer solstice for places on the Meroë-Borysthenes longitude (F24); he placed 15 hours at 'between Naples and Rome' (see on F7). These were based on observation and measurement. As to places further north, there are two points to note. Some of his summer daylight figures are clearly calculated; he could not have taken or accessed actual observations. His 15¼ hours for Byzantium and 16 hours for the latitude through the mouth of Dnieper (Strabo 2.5.41, 42) were from observations, the second (probably) in Olbia. But there was a 15½ hour figure for the middle of the Black Sea, and 17 hours expressed as 6,300 stades (715 miles) north of Byzantium; these were clearly calculated.[5] So when we read his 17 to 19 hour figures for Britain, F28, there is no need to assume that these are, or derived from, Pytheas' figures. Further, he could use 'hour' in the modern sense, which Pytheas could not: see Appendix 3. As noted on F32, it is not easy to see that Pytheas' reading for Marseille was at the summer solstice; if it was, we cannot see it behind the figures in Hipparchos' table(s).

7 We do not know if Hipparchos offered winter sun heights other than those in the north noted on F28. But he could calculate them. The mathematics is simple, but also requires trigonometry. Using a *gnomon* reading of 5:3, for instance, he could say that it corresponded to that ratio of day to night, and use it to establish the latitude of the place: see above. It also meant that the sun's shadow hit the ground at an angle of tan 5/3, with the corollary that it was tan 3/5 at the winter

3 He proposed one table with the rising times of the zodiacal signs, how much of which stars can be seen, the ratios of the *gnomon* to its shadow at solstices and equinoxes, the height of the sun and of the north pole, 'all calculated theoretically'; and a second table with names of places, and their latitude in degrees. Dicks (1960) 193 is a partial reconstruction of Hipparchos, with his latitude for each entry and the actual modern latitude.

4 The question of checking the accuracy of Hipparchos' figures against modern figures is not in point here. Dicks' table, n 3, shows that nearly all his lengths of daylight, there expressed as degrees, were too high by between about half and two degrees. But 19 hours of daylight is in fact correct for the very north of the Shetlands, and substantially so for Mainland. Where his figures erroneously assume that Marseille is on the same latitude as Byzantium, his figures cannot stand up to modern scrutiny because of that error: see on F26.

5 Strabo 2.5.41: 'if you enter the Black Sea and sail north for about 1,400 stades [160 miles, 255 km], the longest day is 15½ hours, and you are halfway between the equator and the north pole'. It is accurate: 160 miles north takes you to about Varna, which we can place at 43° 13′, with a longest day of 15 hrs 23 mins. The 17 hour figure is at *id* 2.5.42. The only thing that Hipparchos knew about land 700-odd miles (c 1,120 km) north of Byzantium, which is in fact further north than Kiev, some 650 miles (1,040 km) distant as the crow flies, was that it was occupied by Scythian tribes.

solstice. Thus even if Pytheas took *gnomon* readings in the north, Hipparchos' winter sun heights are from his calculations, not from Pytheas, and no evidence that Pytheas took such readings. Nor can we say that any of his figures were used by later astronomers. They may have been noted and used as a cross-check, if located with sufficient clarity to make that possible; we cannot go further than that. Note that Strabo presents the winter sun heights in cubits; it is possible that Hipparchos used degrees and Strabo turned them into cubits for the same reason that he turned Hipparchos' latitudes into stade distances, for which see on F28.[6]

8 When we come to Thule, it is clear from F2 and F31 that Pytheas worked out that at Thule the summer tropic and the arctic circle coincided: both probably celestial lines; the sun 'stood still', having reached its highest point or most northerly point on the celestial globe: see on F2 and cf §4 above. Also, as noted on F2, if he referred to constellations, we have lost it. We must now consider what he said about the length of daylight on Thule at the summer solstice.

Our basic references are:

(a) F2 (Cleomedes): Around the island called Thule, where they say . . . Pytheas of Marseille was . . . In this place, when the sun is in Cancer, the day will be a month long, at least if all parts of the Cancer are always visible to them; if not, for as long as the sun is in the parts of it [Cancer] that are always visible.

(b) F15 2.186 (Pliny): Britain . . . [has] continuous days for six months and [continuous] nights . . . at the winter solstice. That this is so in . . . Tyle . . . Pytheas . . . wrote.

(c) F19 4.104 (Pliny): Tyle . . . at the solstice there are no nights . . . conversely no days during the winter solstice. Some consider this is true for six months at a time.

(d) F20 6.219 (Pliny): . . . Thule, in which, as we have said, there are continual days and nights in turn.

F3 and F9 merely copy Pliny and have no independent value. Two other texts need noting:

(e) F13 (Mela): . . . on [Thule] . . . the nights are short. In winter they are as dark as anywhere, and in summer, bright. But at the solstice there are no nights.

(f) Geminos 5.13: If we go even further north, the whole summer tropic is above the Earth, so that at the summer solstice the day at this place is equinoctial 24 hours (trans Evans-Berggren).

---

6 Note that if you check Hipparchos' figures against modern calculations, the summer and winter figures total over 24 hours, because they include twilight as well as night: e.g. Leeds, summer 17 hrs 6 mins, winter 7 hrs 25 mins; Holm of Copister, Shetlands, summer 19 hrs 3 mins, winter 5 hrs 43 mins. Hipparchos had only day and night, the day beginning a little later and ending a little earlier than modern figures offer.

As noted on F2, Cleomedes, text (a), does not derive his month of daylight from anything Pytheas wrote, except indirectly via Eratosthenes. As to Pliny, his bibliographies in *NH* 1 (see on F14) lists Hipparchos, Eratosthenes, and Pytheas for book 2 but only the last two for book 4. Mela, text (d), is a very general statement. Geminos, on the other hand, uses a calculated figure. Text (a) post-dates Pliny, but it is clear that in his time statements about daylight in the north varied from the very general to the scientifically accurate. Also, no text was based on actual readings taken at Thule. Geminos' 24 hours is technically correct, notwithstanding that for several days either side of the solstice, very little of the sun actually sets,[7] and we know that in fact there is a month of daylight on the arctic circle, and two or three weeks of daylight a little further north or south: cf p 155. We cannot attribute such modern knowledge to any of our sources, Pytheas included. Unfortunately, Geminos was not one of Pliny's sources, and Pliny is our only source for deciding what Pytheas wrote; the differences in texts (b) to (d) could be explained if Pliny used different notes for each topic he was writing up (see on F14). It is clear that while astronomers could calculate daylight in the far north with fair accuracy, Pliny did not draw on them. His sources reflected long-standing popular beliefs about daylight and night in the far north.[8]

9 Pytheas would realise that the days were getting longer as he sailed up the west coast of Britain after leaving Cornwall, and shorter in the autumn when along the north coast of Scotland; he would experience long nights when overwintering in the Shetlands; cf the 'sun's bed' reference in F7. But he had no means of measuring the length of daylight (Appendix 3). He lacked both data and mathematics to calculate it, at Thule or indeed anywhere. Unless we adjust his route to put him on Thule at the summer solstice, when he could take a *gnomon* reading at midday and see that there was no or little shadow; or that he stayed for several weeks, when he could measure the continuous daylight by the phases of the moon, his only knowledge of daylight there was what he understood the Shetlanders were telling him; their beliefs based on what they in turn had learnt about Thule from their Norwegian contacts. If that was some generations earlier, it would very possibly be wrongly recalled by Pytheas' time.[9] Thus Pliny is probably accurate in text (b); Pytheas understood six months, and duly recorded it in his book. Roseman (1994) 77 suggests 'correcting' Pliny's text from 'six months' to 'a whole month'.[10] But Pliny's texts have been transmitted correctly, as we see

---

7 Geminos' excursus on lengths of daylight starts at F7 and goes on to 6.42; for the north they are all calculations: at 6.13–15 he has a 24 hour day on the arctic circle, a month further north, two months further north still, and six months at the north pole. Like Cleomedes, above, his calculation for the arctic circle will have partly depended on Eratosthenes' use of Pytheas to establish the arctic circle in our sense (see on F2).

8 Thus Mela 3.36, describing Scythia, says that the Hyperboreans who live beyond the Ripean mountains have six months daylight followed by six of night. Cf Homer's Laestrygonians and Cimmerians, p 10.

9 For instance, as the detail is retold over the generations, 'many days' become 'a long period' and then 'six months'.

10 From 'senis mensibus' to 'toto mense'.

from his confirmation of text (b) in text (d), and her proposal also assumes that Pytheas knew the actual position. As noted in Appendix 6 §16, if Pytheas was ever on Thule, it was a brief stop in early spring. He was not there in the summer; further, he would not be able to communicate in detail with the Germanic speaking inhabitants about their summer solstice experience. Pytheas got it wrong, but through no fault of his.[11]

10 It is reasonable to think that Hipparchos would have included Thule in his table(s) noted above; as we see from the discussion on F28, he could have done the necessary calculation. We rely on Strabo to recover the table(s), as noted in §5. Since Strabo disbelieved in Thule (e.g. in F25), he would not want to copy Hipparchos' figure for it. In theory, Hipparchos' calculation could well be Geminos' source for text (f); though as Pliny used Hipparchos for book 2, it would mean that this detail in Hipparchos got overlooked.

11 The above presents a rather restricted view of what we can infer Pytheas recorded in terms of astronomy; not to downgrade his expertise, or to ignore Strabo's concession in F38 and F39 that he did record some things, but because the material that we have does not enable us to go further without guesswork. It may be contrasted with others' more wide-ranging proposals. Thus Hawkes (1977) 8, 11, 27–8 very specifically proposes that Pytheas took readings at 48° 40′ around Ushant and 54° 14′ at the mouth of the Oder in the Baltic; he refers to both summer and winter readings. Roller (2018) 68–9 says that Pytheas provided all four winter sun heights, expressed in terrestrial cubits ('about 50 cm'): nine cubits for 48° 42′, six for 54° 17′, four for 58° 17′, and less than three for 61° 17′. Dicks (1960) 180 says 'during . . . his travels he made accurate observations of the maximum elevation of the sun in at least three different places', though at 185 he qualifies this: 'the crucial question . . . were the sun-heights and lengths of the longest days actually observed by Pytheas, or were they merely taken [sc by Strabo] from the theoretical data that Hipparchos gave in his astronomical table?' Roseman (1994) 43 notes Dicks' views but doubts that Pytheas could have calculated latitudes without trigonometry.

12 None of such suggestions relate to the realities of an actual voyage, and imply a fairly long period away from home; two or at the most three years is proposed in the Coda, pp 170–7. The difficulty with Roller's suggestion is fourfold. Firstly, to measure the height of a celestial body (in any units) needs a dioptra. They existed in Pytheas' time; they are noted in Euclid's *Phaenomena*, c 300. But it assumes that he foresaw the possibility of needing one before he set off, and

---

11 If we want to attribute more accurate knowledge to him, we must assume that Pliny or one of his secretaries noted Pytheas' text incorrectly. We there enter the realms of speculation. Cf Mela's Hyperboreans, n 7.

included one as part of his luggage.[12] That can, however, be countered by the point noted in §7, that if Pytheas took a *gnomon* reading at a solstice, later astronomers could calculate the winter sun height. Secondly, it assumes that Pytheas spent four winters in the north: one around Brest, which is at 48° 23', and three in England and Scotland; thirdly, that he had the uncanny ability to choose a settlement in just the right area in which to lay up his boat for the winter, so that he could take such precise readings. Fourthly, it would mean that he (and his crew) were away for five years. There is no implication in our references that he was away for so long; if anything, F30 suggests the contrary. When Polybios attacked Pytheas for being too poor to travel, he did so from the standpoint of his own journeys, each of which was within one sailing season; and he complained about the expense (F30 n 2). If Pytheas had been away for five years, we might have expected Polybios to add this to his disbelief. The most sensible comment is Bianchetti (1998) 181: it is possible that some of Hipparchos' figures could ultimately go back to Pytheas ('possano far capo, in definitiva, a P'). We enter the realms of speculation if we go further than that. As a final point, it is doubtful whether astronomy was the main purpose of Pytheas' voyage; it was one of exploration coupled with the investigation of the tin and amber trades. His *gnomon* readings, or any other astronomical observations, were a useful and interesting optional extra.[13]

12 There is the argument from desperation (or imagination) that when Pytheas was in the north, he saw the need to measure the sun's height and constructed his own dioptra on the spot.
13 Apart from the north pole reference, §1, there is little evidence that Pytheas recorded constellations: see p 42.

Appendix 3

THE WORD FOR 'HOUR'

As noted on F7, the Greek word *hora* meant a period of time generally, and was extended to have its modern meaning only when sundials were developed and a word was needed for the 24 subdivisions of the day. The evidence shows that in Pytheas' time the word did not have that meaning. No sundials earlier than the third century are known (Gibbs (1976) 5), but that means only that any earlier devices have not survived. Literary references to earlier devices, conventionally translated 'sundial', use different words, and some only showed the solstices and equinoxes.[1]

Anaximander of Miletos, early sixth century, is said to have invented four items; this is in Diogenes Laertius' *Lives of the Philosophers* 2.1, a book of the third century AD. It is basically second or third hand material from earlier writers, and so not always reliable. One item, a map, confuses him with his pupil Anaximenes, for which see Dilke (1998) 22–4. The second, a globe (σφαῖρα, *sphaira*), could be accurate if it refers to the celestial sphere.[2] The third is almost certainly wrong; he is said to have invented the *gnomon* (Appendix 2 §3), which he used on a σκιοθήρης, *skiothērēs*, 'shadow chaser', several of which he erected in Sparta; it showed solstices and equinoxes. The Ionians absorbed much astronomy from Babylonia and Egypt (so, e.g. Guthrie (1962) 33–4), but it is very doubtful if Anaximander could establish solstices and equinoxes: Dicks (1966); briefly *id* (1970) 45. If he could, his device was not a sundial.[3] The fourth item is possible. It is said that he constructed ὡροσκόπια, *hōroskopia*, literally 'hour-lookers'. The

1 For instance, that on Chios fully described in Hunt (1940–45) 41–2. The problem is briefly noted in Bowen and Goldstein (1991) 239–40.
2 But not if it represented the earth: Anaxagoras, two generations younger than Anaximander, thought that the earth was flat, though around his time the Pythagoreans concluded it was round: Dicks (1970) 58, 72, and cf p 20.
3 Gibbs (1976) 5–7 is helpful on the history of the sundial, but her mention of Anaximander must be read subject to Dicks' caveats. Pliny *NH* 2.187 attributes the device to Anaximander's pupil Anaximenes. A troubling detail is that although Vitruvius used the Greek word *sciatheres* to mean sundial, in Greek texts it meant a *gnomon*: Vitr *De Arch* 1.6.6; Gem 6.43; but the context in Diogenes shows that he (Diogenes) means a device of some sort, not the *gnomon*.

   DOI: 10.4324/9781003181392-7

word usually had an astrological meaning, but could in Diogenes' time refer to a sundial. In view of the mention of the *gnomon*, it is possible that Anaximander built a version of the Egyptian device described by Herodotos in the passage next noted. If so, Diogenes' use of *hōroscopion*[4] is no evidence that Anaximander used *hora* for its subdivisions, nor how many subdivisions it showed.

Herodotos (mid fifth century), within his description of Egypt, tells how Egyptians survey plots of land, adding that this is how geometry (his word as well as ours) reached Greece, and continues 'however the Greeks learnt about the πόλος, *polos*, and γνώμων, *gnōmōn*, and the twelve parts of the day from the Babylonians'.[5] The *gnomon* is the vertical rod to catch the sun's shadow. *Polos* had several meanings, including the celestial sphere or sky or the celestial pole in F8; here it probably means a bowl of that shape, i.e. hemispherical, as many ancient sundials were.[6] The significance is that Herodotos does not know 'hours'; he says that the day and night are divided into 12 μέρεα, *merea*, 'parts'.[7] It is one thing for Greeks to have learnt about such devices; it is another to assess how common they were. Aristophanes made a humorous reference to this *polos*: his character asks, by how much has the sun moved?[8] As with all such lines, it is hard to know if they are an exaggerated take on something usual, or a shot at a seven-day wonder. His 'how much', *postēn*, is a feminine adjective in the accusative, and Greek could use such adjectives where the noun was understood, e.g. Hdt 6.21.1; but we cannot use this passage to infer that the missing noun was *hora*. It is equally likely to have been (for instance) χώρα, *chōra*, 'space', or διάστασις, *diastasis*, 'distance'. It is known that the astronomer Meton erected a *heliotropion* ('sun-turning [thing]') in Athens some time before 427, which showed the solstices and equinoxes, and Aristophanes made a humorous allusion to it, using again *polos*.[9] After the fifth century, *polos* disappears from the literary record for any type of sundial, though

4 For the eventual meanings of the word, see *LSJ* svv ὡροσκοπεῖον, ὡροσκόπιον, and ὡροσκόπος.
5 Hdt 2.109.3. That the passage is genuine and not an interpolation was settled by Robertson (1940), though his very helpful review of the evidence does not distinguish between sundials and devices to mark solstices and equinoxes.
6 Gibbs (1976) 4 reports that no horizontal disk-shaped sundial is known; the usual shape was spherical or conical.
7 Note also that Herodotos came from Ionia, the same area as Anaximenes and Anaximander. He would have known if either had developed devices called *hōroskopia* for marking divisions of the day which they called hours.
8 Ar F169, from his *Gerytades*: A πόλος τόδ' ἐστί; B κᾷτα πόστην ἥλιος τετράπται; (A Is this a *polos*? B Then by how much has the sun moved (literally 'turned')?). B's question begins with the sarcastic κᾷτα, καὶ εἶτα, *kai eita*, 'and then'; the natural meaning is that Aristophanes is poking fun at something new and unusual; Henderson in his Loeb Fragments cannot be right with 'probably the pole of the underworld, as opposed to the celestial pole'. The grammarian Pollux 9.46 preserves this fragment, noting that *polos*, as in Aristophanes' lines, was perhaps (που, *pou*, 'possibly') equivalent to *horologion*, 'hour reckoner or calculator', which was (and still is) the usual word for a sundial.
9 Philochor *FGrH* 328 F122; Ar *Banqueters* F227; useful discussion at Dunbar (1995) 554–5. Aristophanes jokes that it looks for *meteōra*, 'things in the air' and *ta plagia*, 'sideways things'.

it appears to have remained in the colloquial language as a shorter word for the literary *hōrologion* and *hōroskopion*, possibly as a feminine noun.[10]

Significantly, when Vitruvius sets out the history of sundials, he begins with Eudoxos, a generation senior to Pytheas and the source of much of the latter's astronomical knowledge (pp 19–22). At *De Arch* 9.8.1, having explained the correct way to mark out a sundial, Vitruvius names the types and the men who devised them. They are all third century or later, except for this: 'the Arachne ('Spider' or 'Spider's Web'), by the astronomer Eudoxus or, as some say, by Apollonius' (the mathematician Apollonios of Perga, c 260–c 190). However, Goldstein and Bowen (1983) 333, 335–6 are clear that it was Eudoxos, not Apollonios.[11]

We have no actual evidence from which we can infer how he marked his sundial. Vitruvius implies that it showed more than equinoxes and solstices, and so we can infer that it marked parts of the day. However, we do not know what he called those divisions.[12] If it was *hora*, that use of the word would be known only to astronomers who were privy to his work. It is clear that when describing the length of the longest day at the summer solstice, he did not express it as so many hours, but as a measurement, the ratio of the lengths of day and night, corresponding to the ratio of the lengths of his *gnomon* and the sun's shadow (p 20).

Our earliest evidence for *hora* as 'hour' is P Hib 27, a calendar for a year between 301 and 298 from an astronomer to his pupil (Grenfell and Hunt (1906) 138–9, 155); and Timocharis of Alexandria, same period, used *hora* in observations preserved in Ptolemy (Bowen and Goldstein (1991) 240). Our earliest non-scientific use of the word is in the comic poet Baton, c 250. He makes fun of a parsimonious teetotal philosopher, saying 'the man is too mean to buy wine, and he looks at his (empty) oil flask (*lēkythos*) from morning to night to see if there is still a drop in it; you would think that he was carrying a *hōrologion*, not an oil-flask'. The device was not a water-clock; it was a new-fangled sundial, but sufficiently common for the audience to understand that it was marked in hours in our sense. Baton was three generations after Pytheas; the modern sense of 'hour' is common from Hipparchos

10 It is feminine at Ar F227, and also in the second century AD satirist Lucian, *Lex* 4, who describes the time when the *gnomon* casts a shadow on the *polos*.
11 Apollonios may have improved earlier types of sundial, but they must have existed before our earliest survivals, and he was born in the second half of the third century (Toomer (1970) 179–80). He was an important mathematician, and whatever he did with sundials could not have been before the last quarter of the third century. For sundials generally, see Gibbs (1976) 4–5, 10.
12 Nor we do know if he envisaged a 24 hour day, though it is likely that he did. Note that on sundials Greeks typically used their alphabet: α for 1, β for 2, etc; so, e.g., the sundials 1001 and 1044 shown in Gibbs (1967) 122–3, 158–9. For their numbering system generally, see *OCD* sv 'numbers'.

(e.g. in his *Commentary*, from which our F8 comes, at 2.4.5) onwards.[13] Latin does not help: the sundial and *hora* for an hour were established at the time of our earliest surviving text, Plautus, in the late third century (F7 n 6).

Thus, as briefly suggested on F7, Pytheas could only have used *hora* to mean 'hour' in our sense if (a) that was the word Eudoxos used for his divisions; (b) he devised a 24 hour day; and (c) such sundials were already so common that the usage had already passed from astronomical writings into common speech, and his readership would have understood what he meant by the word. We cannot infer that on our evidence, and we should bear in mind that Greeks did not have our obsession with time, much less clocks and watches or a universal standard for measuring time; cf the comments of Dicks (1966) 29. For a long time, the public would regard a sundial as an astronomer's curiosity, not something of general utility.[14]

The above identifies how *hora* came be used in its modern sense. It is a separate, though related point, to the distinction between equinoctial hours (F2, F7), the time shown on sundials that were calibrated at an equinox so that the hours were of equal length throughout the year, and ὧραι καιρικαί, *horai kairikai*, seasonal hours, $\frac{1}{12}$ of daytime, which varied throughout the year; sundials so marked were accurate only for the place where they were set up. The papyrus hours were (probably) equinoctial; Timocharis' were not, and that was probably the case with those with which Baton's audience were familiar.[15]

---

13 Baton *Murderer* F2. There was a dedicated word for water-clock, *clepshydra*, 'water-stealer', and it gave only the period of time that elapsed until its container of water drained dry. The *hora* in *hōrologion* must mean an hour in our sense. We should imagine either a stone sundial, where part of the joke would be the difficulty of carrying a heavy item with you, or a more portable one made of copper or brass. Either way, the joke refers to a novel and so unusual item. At this early date, it is very unlikely that there were already portable sundials, 'viatorium pensile', 'travelling [item] capable of being hung up', noted by Vitruvius *De Arch* 9.8.2.

14 Dicks (1966) 29–30, (1970) 165–6, grasps this point: the *gnomon*, i.e. a sundial able to denote hours, was not used for measuring time until Hellenistic times.

15 For the distinction see Bowen and Goldstein (1991) 239; Gibbs (1967) 10. Although the papyrus has 'hours' for most time, for the solstice the word is δωδεκατημόρος, *dōdekatēmoros*, 'twelfth part'; the passage is complicated because of other figures: see Grenfell and Hunt (1906) 155 on lines 116–22. Once common, ordinary sundials were calibrated for the latitude of the place where they were made (Vitr *De Arch* 9.7.1–2). In the Roman world, daylight was always 12 hours, so that the length of an hour varied through the year: see Gibbs (1967) 4–5. Pliny *NH* 7.214 has an anecdote of a sundial taken to Rome from Sicily in 263 and used for 99 years without it being realised that it was inaccurate for Rome. At *NH* 2.182, just before F14, he notes that ordinary sundials are not accurate everywhere (beware of Rackham's Loeb translation 'travellers' sundials'; Pliny is speaking generally).

Appendix 4

PYTHEAS' AMBER ISLAND(S), WITH A NOTE ON GERMANY AND SCYTHIA

Several persons, including Pytheas, are named by Pliny and Diodoros as referring to an island where amber is found, but with no agreement as to its name. Two different names are ascribed to Pytheas. The references cannot be reconciled logically unless we make assumptions, but they all require to be taken into account if we are to decide what Pytheas wrote, and where his island(s) was or were. It is convenient first to set out the texts.

(i) Pliny *NH* 4.94–6 (from F17), describing the Baltic:

[94] *Many unnamed islands in that area are reported. Timaios reported one called Baunonia, a day's sail from Scythia, where amber is cast up by the waves in spring . . .* [95] *. . . Xenophon of Lampsacus reports a very large island, Balcia, a three days' voyage from the coast of the Scythians. Pytheas calls this Basilia.* [96] *. . . the Inguaeones clan [are] the first in Germany . . . [there is] the huge Codanus Bay by the promontory of the Cimbri, which is full of islands, of which the most famous is Scatinavia, of unknown size.*

(ii) Pliny *NH* 37.35–6 (from F21), within his discussion of amber:

[35] *Pytheas [said] that there is a tidal coast occupied by the Guiones, a Germanic people, called Metuonis, 6,000 stades [680 miles, 1,090 km] long. From here, the island Abalus is a day's voyage away. It [amber] is carried to it in spring by the waves; it is a discharge from the frozen sea. The inhabitants . . . sell it to their neighbours, the Teutoni.* [36] *Timaios also thought this, but called the island Basilia.*

(iii) Diodoros 5.23.1, 4 ≈ Timaios *FGrH* 164.23 (from F6), dealing with electron, i.e. amber:

[1] *Opposite the part of Scythia which is beyond Galatia [Gaul], there is an island in the Ocean called Basilia. On it the waves throw up . . . electron in great quantities. . . .* [4] *. . . amber is collected on the island . . . and it is taken by the locals to the mainland opposite.*

DOI: 10.4324/9781003181392-8

Island names: overview and preliminary analysis

1 (a) Whatever Pytheas said was based on autopsy. Any other author was either using Pytheas, or recording how merchants at the Mediterranean end of trading routes spoke of the origin of amber.

(b) As noted on F6 and F21, most amber came from the mainland, so when merchants spoke of their amber coming from an island, it could at best be only half true. Just as tin merchants said that tin came from an imaginary archipelago, merchants were claiming a distant island origin for their amber: the further away its alleged origins, the easier to justify its cost. Although most if not all the amber reaching the Mediterranean will have come from the mainland, we cannot infer from our evidence that any of the names referred to a place on the mainland. The merchants spoke of it coming from an offshore island. We should, however, be alive to the likelihood that the way they pronounced a given name would be different to how Pytheas heard the name when on the spot.

2 As to what Pytheas wrote, we must resolve the following:

(a) Assuming that Timaios drew on Pytheas, text (iii) suggests that Pytheas wrote 'Basilia', apparently confirmed by text (i); but Timaios also had 'Basilia', text (ii).[1] But text (ii) says that Pytheas had 'Abalus'. In both cases, the amber is cast up by waves; but text (ii) adds two details: it is off the Metuonis coast, and it happens in the spring: see §7 below. We could partially reconcile the names if we assume that 'Abalus' is from a misreading in Pliny's notes; for his gem section, book 37, he would be using a different set of notes from those he had for his geography of the north, book 4 (see on F14 for his method of working); and the Metuonis coast and frozen sea details in Pytheas' original were either not copied by Timaios or omitted in Diodoros' epitome.

(b) An alternative view is that Pytheas found two amber islands, one placed generally off 'Scythia' and one more precisely off the Metuonis coast.

(c) But text (i) attributes 'Baunonia' to Timaios; apart from the initial B, it is not close to the other names, though the details of waves and spring are identical with the waves in text (iii) and waves and spring in text (ii). The simplest solutions are to assume either (a) 'Baunonia' is an error in Pliny's notes, and should be 'Basilia', as recorded in text (iii) and so from Pytheas; or (b) that Timaios named two islands, Basilia from Pytheas and Baunonia, which he heard from amber merchants when he lived in Athens, omitted in Diodoros' epitome.

---

1 Pliny would not realise that Timaios drew on Pytheas. He was used to reading the same thing in different authors, not appreciating that one had copied the other; cf treating Isidoros as independent of Pytheas, F17.

(d) The problem with Xenophon's Balcia (Βαλκία, *Balkia*, in Xenophon) is two-fold. One is that Pliny was apparently not troubled by the different pronunciations of the middle consonants, Xenophon's -lk- as opposed to Pytheas' -s-l-. He could be right, for reasons that he would not appreciate, if we assume that the local pronunciation was (say) 'Balchia', with 'ch' as in 'loch', or 'Baltsia' or 'Baltchia', which Pytheas rendered euphoniously as 'Basilia' and Xenophon's amber merchant informants pronounced 'Balkia'. But he appears to have overlooked the second point, that this island is three days from the mainland. Text (iii) does not imply so distant an island, and all the other amber islands are specifically within a day's sail. This can be resolved in three ways: (1) it is a different island from Pytheas', being three days from the mainland: Öland or Gotland to the Polish coast would suit; (2) the distance was exaggerated in the mouth of Xenophon's informants, and they were referring to the same island as Pytheas; (3) they were speaking of an imaginary island.

(e) The balance of probabilities suggests that Pytheas wrote 'Basilia' and Timaios copied him; 'Abalus' results from poor notes, and Pytheas only named one island. If Balkia and Baunonia are not this island, locating them can only be guesswork.

Locating Metuonis and Pytheas' island

3 As a preliminary comment, none of the islands, nor Metuonis, are in Ptolemy. In the period between Pliny and Ptolemy, it is hard to find evidence of any direct interest in the area which could have given Ptolemy the data for including them. See F17 n 1 for the different approaches of the two men.

4 We have two clues for locating Metuonis: its size, and Pliny describing it as an 'aestuarium'. As to size, it is one of the few places where Pliny did not translate a Greek author's stades into Roman miles. Also, it is difficult to see that the number, 6,000, has got misread in transmission, or that Pliny's note-takers would get it wrong.[2] It seems safe to assume that it is Pytheas' own figure, and that it reflects his method of turning days' sailings into land distances. While we should treat it with caution, we should assume that he intended to speak of a substantial distance of several hundred miles.

5 What did Pytheas write that Pliny translated as 'aestuarium'? The *OLD* translates (1) an inlet, etc, covered by the sea at high tide, tidal opening; (2) a river estuary. It cannot be the latter; no estuary anywhere is even 300 miles

---

2 But Bianchetti (1998) 196 queries whether Pytheas' number was lower and got misread by Pliny. The actual shape of the coasts both west and especially east of Jutland are a complicating factor: how close did he cling to the shore? Did he sail at least partly round some of the islands?

long.[3] Nor are there inlets of such dimensions. It is here translated 'tidal coast'. There is no precise comparable for this meaning in surviving literature, but it corresponds to the reality of the two possible locations for Metuonis, west or east of Jutland.[4] Greek did not have words either for estuary[5] or a tidal coast; Pytheas must have used a periphrasis, though that does not help in locating it. Tidal coasts are essentially absent in the Mediterranean and did not need a word. He perhaps used a phrase similar to that in Ps-Arist *Marvels* 844a 24–30, where a coast west of Cadiz is described as covered in rushes and seaweed at low tide but submerged at high tide, where large tuna were harvested ('pickled, packed in jars, and sent to Carthage'),[6] or alternatively a phrase using παλίρροια, *palirrhoia* (or the equivalent verb), a flowing back and forth. Pliny would see that his 'aestuarium' was a suitable equivalent. Pytheas did not mean a bay: for that he had κόλπος, *kolpos*, and Pliny would have translated 'sinus', the Latin equivalent.

6 There are three candidates for Metuonis: (1) the Wadden Sea, the mud flats and salt marshes along the north coasts of Holland and Germany and the south-west coast of Jutland. It is mostly a UNESCO World Heritage site some 310 miles long; (2) the tidal islands around Denmark discussed on F13 §55; (3) the long stretches of tidal beaches along the Baltic coasts of Germany and Poland.

7 Three points tip the balance strongly in favour of the Baltic. One is the mention of spring waves and discharge from the frozen sea, text (ii). As noted on F21, in spring the tides are swollen with melt-water from the Gulfs of Bothnia and Finland. The locals would know from trading contacts about this frozen sea which melted in spring. Pytheas could not, unless the locals told him so: he learnt it on the spot. He was strictly wrong about spring; amber is cast up all the time, but there would be more in spring. He would naturally assume that the locals were speaking of the same frozen sea as that north of Thule; he could not know that the Baltic was a cul-de-sac. If we identified Metuonis with the Wadden Sea, he would be on, say, Heligoland or Sylt. Certainly amber from the west coast of Jutland would wash up there, but the locals could not know of a frozen sea in the distant north; nor would their mainland contacts, on, say, the German

---

3 There are ordinary-sized estuaries in the relevant areas: the mouths of the Weser and Ems, and at Kiel, for instance. As a footnote, the large Amazon estuary is about 150 miles wide and about 230–250 miles from the sea until the water is narrow enough to be thought of as a river.
4 We must not be influenced by its derivative 'estuary' in various modern languages.
5 When Strabo and Ptolemy needed to describe an estuary, they had to invent and explain. Strabo used ἀνάχυσις, *anachysis*, literally a 'pouring up', with an explanation as to its meaning, e.g. 3.1.9: 'hollows that are covered by the sea at the high tides, and, like rivers, afford waterways to the interior', (trans Jones); it is unknown in this sense before him. Ptolemy used εἴσχυσις, *eischysis*, 'pouring in', again more or less unknown outside his works.
6 The Greek translated 'submerged' is from βαπτίζω, *baptizō*, hence its eventual use for the ceremony.

coast west and north of Cuxhaven.[7] Pytheas' information must have come from those with knowledge of a frozen sea, i.e. the Baltic, and so those living east of Jutland. Philemon was to learn the same thing from some source over 300 years later: see F17.

8 Secondly, also in text (ii), Metuonis is home to the 'Guiones', and the islanders deal with the Teutoni. As discussed in the appended Note (see §§9 and 11), it is probable that the 'Guiones' are the Inguaeones, and Pytheas' island has to be within a day's sail of the east side of Jutland. We should also note Strabo's references to the 'lies' Pytheas told about Scythia, for him somewhere east of the Elbe. As pointed out on F25 §3, this almost certainly means the Baltic, and see also on F39. The third point relates to Pliny's 'aestuarium'. It is tempting to suggest that Pytheas described the tides round the islands east of Jutland, modern Denmark, in some detail, and that some of this is preserved in Mela's description of tidal islands in F13: see particularly the reference on F13 to Svennung (1963); while Pliny merely summarised Pytheas with the one word 'aestuarium'. As noted on F13, whatever Mela thought, there is in fact only one group of such islands.

9 On that basis, we may eliminate (1), the Wadden sea. As to (3), the tidal beaches east of Rostock, we could account for Pytheas' 6,000 stades if he included a stretch of this coast in his estimate of distance. Given that Metuonis is east of Jutland, where does it lie in relation to the Codanus bay? We may reasonably infer two things: (a) Pytheas described the Metuonis coast too loosely to enable Pliny to include it in his geography of north Holland and Germany, basically F17 and what follows; and (b) Pytheas did not describe any part of those coasts in such a way as to enable Pliny to recognise Tastris, Jutland. Whatever Pytheas said, the coast was little known until the extent of Jutland became known from the expedition of 5 AD referred to on F17. We have no evidence that Eratosthenes included the Baltic (or the rivers of northern Europe) in his *Geography* (Roller (2010) 29), and while it is dangerous to infer too much where a work survives only in fragments, it does suggest that Pytheas said too little about the Baltic to be of use to Eratosthenes. Posidonios reported a peninsula where the Cimbri lived, but with no sense that it was of any great size; the context in which Strabo refers to it is to the same effect.[8] When reports of the Codanus Bay first reached the Roman world, it was not located in relation to a peninsula, as we see from Mela, texts (v) and (vi) in the Note.

---

7 By ignoring that Thule is six days north of Britain (F24), and the frozen sea another day beyond (F19, F24), some German scholars identify the frozen sea with muddy shallows of Jutland: see Roseman (1994) 130; Bianchetti (1998) 166. Note that, with lower sea levels, Heligoland may have been one island in Pytheas' time. It is some 40 plus miles from the west coast of Jutland or the Cuxhaven area, 50 plus miles to Wilhelmshaven and Bremerhaven. Although the North Sea can be rough, they must have had boats sturdy enough to maintain contact with the mainland.

8 Posidonios F272 E–K = Strabo 7.2.1–2; F49 E–K = Strabo 2.3.6.

Conclusions

10 Where an island is said to be off 'Scythia', in a text going back to an era before Germany was known to exist between Keltikē and Scythia, we should understand it to be the Baltic.

11 The Metuonis coast is east of Jutland. The Teutoni who bought the amber lived in Jutland and were in fact within the clan Inguaeones; it is the latter whom Pytheas referred to as living on this coast and who appear in text (ii) as 'Guiones'. Although we cannot be sure how much of the coast Pytheas meant with his 6,000 stades figure, the realities of what is actually in the area mean that we should think of Metuonis as broadly between Flensburg and Rostock, and perhaps a little further east, with the Teutoni living closer to Flensburg than Rostock.

12 This makes one of the southern Danish islands a strong candidate for Pytheas' Basilia, an identification not affected by the likelihood that they are also Mela's tidal islands discussed on F13 §55. Rügen and Bornholm would also probably be opposite Inguaeones territory; Rügen cannot be ruled out, but the distance of Bornholm from the mainland makes it less likely. This identification does not mean that Pytheas did not sail further east; it is possible that he went further east into the Baltic before turning back. But, as noted on F13, beyond Rügen there are no more islands, and he never found the real source of amber, Samland. Either Balkia or Baunonia could be the same island; if not, the possibilities are canvassed in 2(c) to (e) above.

13 There is one final point. At *NH* 37.61, in his amber section, Pliny quotes one Metrodoros of Scepsis, c 145–70, for an amber island off 'Germany' which is spelled 'Balista' in Pliny's MSS. Usually amended to 'Basilia', 'Balkia' is also proposed (see Bianchetti (1998) 204)). For all we know, 'Balista' could be how Metrodoros heard the name, but his island can be fitted in to the above analysis by identifying it with Pytheas' Basilia.

14 The above analysis is based on the specific assumptions made here; it not surprising that others have assessed them in different ways, sometimes with considerable ingenuity.[9]

---

9 Thus Hawkes (1967) 9, ignoring Pliny's Abalus = Basilia, proposes that Abalus means 'Apple Island', noting German *apf-*, Celtic *aball*; others have compared Abalus to Avalon and an island of honey in Celtic mythology (see Bianchetti (1998) 196). Bianchetti devotes several pages to the diverse views about all these islands and Metuonis at 195–200 on her F15, our F21, text (ii), and 200–205 on her F16, our F17, text (i). Barrington 10 tentatively locates the Metuonis coast west of Jutland, with Abalus as Heligoland. Zehnacker and Silberman (2015) 319–20, after a brief discussion, conclude that they all refer to one island, or islands near each other. Dion (1966) 201–2 with map fig 11 at 207 imaginatively proposes that Pytheas reached a point on the coast of Kaliningrad just south of its frontier with Lithuania, and called it 'Tanais' to mark the frontier between Europe and Asia in the north, since the Tanais in the Black Sea became known as where Europe ended and

Note: Germany and German tribes, and Scythia

Both these topics are relevant to locating places and tribes in several of our references, particularly Pytheas' amber island and whether the Ost- tribe in F25 is in Germany. Scythia was always little known in areas distant from the Black Sea; we are concerned where in northern Europe its western limits were thought to be. But northern Europe was also little known: the Rhine and German tribes became known after Pytheas' time. For him, Scythia probably began somewhere beyond Keltikē, wherever the eastern limits of Keltikē were thought to be (p 10). Since oral knowledge precedes a written record, that the Rhine existed, and German tribes lived to its east, probably started to become known in the Mediterranean during the second century; Scythia moved east and eventually its western reaches began at the Vistula.

1 Our earliest reference to Germany is Posidonios, a generation before Diodoros, who noted a German custom.[10] He is also one of our sources for the aggressive migration of parts of the Cimbri and Teutoni tribes into Europe, c 114–101; but they are always 'Cimbri', not 'German'.[11] He may have reported the Rhine as a tidal river.[12] Diodoros mentions Cimbri (n 11) but not Germans; he knows of the Rhine: when Caesar bridged it, it was to attack the 'Celts' on the other bank.[13] Beyond the Celts and Gaul it is Scythia: 5.32.1.

2 Caesar does know that German tribes live east of the Rhine and Gallic tribes to the west, noting that some crossed the Rhine when forced out of their homeland

Asia began in the south, probably already in Pytheas' time (Ps-Scyl 68.23, 70.2). Hawkes *op cit* 6–8 more modestly takes him only to the mouth of the Oder at Stettin/Sczeczyn. In any list of Danish islands, note that in Pytheas' time, and indeed until 1825, when the North Sea broke through, the northern part of Jutland was not a separate island; Limfjord was a fjord landlocked at its western end: F17 n 1.

10 Germans eat roast joints and drink milk and neat wine: F73 E–K = Athen iv 153e; F277b E–K = Eustath on Hom *Il* 13.6. The identity of these Germans is discussed in Kidd (1988) 323–6 and Theiler (1986) II 111.

11 Sources: Greenidge and Clay (1960) 61–4, 72, 81–4, 88, 93–4, 97–8, 103–4. Posidonios says that the Cimbri lived on a peninsula (Strabo 2.3.6 = F49 E–K) and are 'plunderers and vagabonds' (Strabo 7.2.2 = F272 E–K); so Diodoros 5.32.4; Livy *Perioch* 63 calls them 'gens vaga', a wandering tribe. Plutarch *Mar* 11.3, 200 years later, in the context of Marius battling the Cimbri, says that the Germans call them 'the plunderers'.

12 The Posidonios reference is late (Priscianus Lydus, *Solutiones* VI 72.2–12, 6th century AD); it couples the Rhine with the statement that the Thames' tides last for four days: Posidonios F219 E–K; Kidd (1988) 786–7.

13 Our earliest certain references to the Rhine are Diodoros 5.25.4: the Rhine and the Danube flow into the Ocean (the Danube mention is from a mistaken belief that it was not the same as the Ister, the Greek name for the Danube); Caesar *BG* 1.31, dealing with the events of 71; and Ps-Arist *Mirab* 846b 29–30, of about the same date, that the Germans live beyond the Rhine and the 'Paeonians' beyond the Ister: 'Paeonians' is probably an error, perhaps of transmission, for Pannonia, roughly modern Hungary.

(e.g. *BG* 2.4, 4.4.3–4, 6.2). But his knowledge of German tribes is limited to those close to the Rhine with whom he came into contact.

3 By Strabo's time, more was known from contact with German tribes. The land west of the Rhine was under Roman occupation, although not formally German provinces until 85 AD, after Pliny's death.[14] From 12 BC the Romans crossed the Rhine and sought control up to the Elbe, though the Teutoburg forest defeat in 9 AD led to gradual withdrawal, completed in 16 AD, with a political decision to keep Roman Germany west of the Rhine in the north. Strabo mentions Germany several times in his introductory sections, books 1 and 2, e.g. the mouths of the Rhine, F25; his description is modest compared to, say, France or Spain, but he names a large number of tribes and gives other details for the area between the Rhine and the Elbe, 7.1.[15] Book 7.2.1–3 deals with the Cimbri. He repeats that they live in a peninsula (from Posidonios, §1 above), though without any suggestion that it is large, and he speaks about their customs. Then at 7.2.4 he says that German tribes extend to the Elbe, including the Cimbri; but the land beyond the Elbe facing the Ocean, i.e. in the north, is unknown. His Cimbri, and their peninsula, are west of the Elbe. He does, however, note tribes living east of the upper reaches of the Elbe, including the Hermondori.[16] He would know of the expedition round Jutland of 5 AD noted on F17, but presumably had already written his German section and did not revise it.[17]

4 Mela was writing a generation or so later, around 40–45 AD. According to him:

(iv) 3.25, 33, 36: [25] *Germany extends [from the Rhine] to the east up to the boundary with the Sarmatians, where its coast faces north and the Ocean. . . .* [33] *Sarmatia is separated from the land beyond by the Vistula. . . .* [36] *From there [the Vistula] Scythian people live.*

(v) 3.31–2, concluding his description of Germany, and having just mentioned its rivers, ending with the Albis (Elbe): [31] *Beyond the Albis, the huge Codanus Bay is full of large and small islands. . . .* [32] *On the bay are the Cimbri and the Teutoni; farther on, the farthest people of Germany, the Hermiones.*

---

14 Germania Inferior in the north, the area between Tongres in Belgium and Cologne, and up to the North Sea; Germania Superior in the south from Remagen into Switzerland. The Romans had difficulty in distinguishing some of the tribes in these areas as between Germans and Celts. Superior later extended across the Rhine.

15 One tribe is the 'Boutones' (7.1.3); otherwise unknown, it could be a mishearing or misspelling for the 'Goutones', Goths. Pliny places them correctly west of the Vistula, §6, but it is possible that some tribal names circulated orally without much clear idea as to their locations.

16 Tribes east of the Elbe: 7.1.3 Hermondori, Langobardi; 7.3.1 Suebi.

17 As to the date of his writing, at 7.1.4 he refers to an instruction by Augustus that the army should not cross the Elbe, showing that he was writing while the Romans were trying to control Germany up to that river, and before the Teutoberg disaster led to making the Rhine the frontier.

(vi) From F13, within his description of islands: . . . *in what we have called Codanus Bay, one [of the islands] is Scadinavia, which the Teutoni still hold. . . . The islands opposite the Sarmatians, on account of the ebb and flow of the tide . . . seem to be islands at some times and other times land.*

His ignorance of Jutland and his belief that the Codanus Bay begins at the Elbe, and his other perceptions of the Baltic, are summarised on F13. But he knows more than Strabo; his Hermiones and the Sarmatians live there, and he is the first to record the Vistula: Scythia now begins there. Perhaps this river had only recently become known. The Teutoni also live in southern Sweden, Scadinavia; but he names only three German tribes: Cimbri, Teutoni, and Hermiones.

5 Pliny has much more detail than any earlier writers. In addition to text (i), we have:

(vii) *NH* 4.99–100 [99] *There are five German clans: the Vandals, who include . . . the Gutones; . . . the Inguaeones, who include the Cimbri, Teutoni and the . . . Chatti; [100] near the Rhine the Istuaeones . . . ; inland the Hermiones . . . ; the fifth part the Peucini. . . . Notable rivers that flow into the Ocean are . . . the Vistula . . . (trans Rackham, adapted).*

(viii) *NH* 3.97, from F17 *Some report that this area up to the Vistula is inhabited by Sarmati, Venedi, Sciri and Hirri.*

He is clear that Jutland is a substantial peninsula, at the western end of the Codanus bay, with the 'islands' of Scatanavia and Aenignia north of the mainland. But, as noted on F17, he is inconsistent as to which German clan you first meet going from east to west: the Inguaeones, text (i), or the Vandals, text (vii). It might be an oversight, e.g. if his notes for (i) were for islands and other details for the Codanus bay area, and those for (vii) listed Germanic clans and tribes. Although (vii) comes only a couple of pages after (i), the discrepancy eluded him; perhaps because with text (i) he had to accommodate text (viii). The tribes in text (viii) are discussed on F17; they are relevant here because, whatever the reality, there was some uncertainty as to where German tribes ended and non-German tribes began in the areas west of the Vistula, and Pliny could be no clearer on this than his sources.

6 About a generation after Pliny, we have Tacitus' *Germania*. He has just three clans, named after the three sons of the divine or half-divine founder of their race: Ingaevones (so spelt) nearest the Ocean, Herminones in the middle, and the Istaevones: Pliny's second, fourth and third clan. He also names many individual tribes, and these include the Cimbri in Jutland, the Varini in the south of Jutland (Pliny names them as a Vandal tribe), the Vandalici, a variant spelling for Vandals, unlocated but probably, as with Pliny, in the east, and the Got(h)ones, a spelling for the Goths that became standard.[18]

---

18 For the three clans and the Vandilici, *Germ* 2; Cimbri, *ibid* 37; Varini, *ibid* 40; Goths, *ibid* 44. He notes the Hermunduri (Strabo's Hermondori, n 16) as living towards the south 'where the source of the

Overview

7 This note does not list all the Germanic tribal names known to us, and is limited to those of immediate relevance.[19] Any wider discussion would have to take into account the possibility of voluntary or forced migration accounting for differences between earlier and later sources.

8 Diodoros had little knowledge of what lay beyond the Rhine. Pliny did know that Scythia began at the Vistula, but does not consider if the island in text (i) is so far east. We may interpret both texts on the basis of what is actually there, and treat them as referring to the Baltic west of the Vistula.

9 Despite Strabo's and Mela's ignorance of Jutland as a large peninsula, it is clear from how they describe the homeland of the Cimbri (and of the Teutoni in Mela), coupled with text (i), that the Cimbri still lived there, and from Mela, text (v), that some lived on parts of the coast between Flensburg and Rostock.

10 Pliny's grouping Germanic tribes into five clans (*gentes*) is neat, and while it reflects Roman perception of how tribes were grouped, for our purposes it can be taken as broadly accurate in terms of where tribes lived.[20] Moving from east to west, once you crossed the Vistula, you would meet German tribes. While Mela's sources reflected less up to date knowledge than those used by Pliny, and were wrong about where the Hermiones lived,[21] his mention of Sarmatians west of the Vistula leaves open the possibility that some non-Germanic tribes lived west of the Vistula to some extent.

11 Finally, we must deal with Pytheas' 'Guiones', text (ii). The question is not whether they are the first tribe or clan you come to in Germany, but whether we

Elbe is', *ibid* 41–2, and taken with Pliny's location of their clan, the Hermi(n)ones, it supports the view that Mela's location by the Ocean is an error. Ptolemy had Goths in southern Sweden: *Geog* 2.11.16, the Γοῦτοι, Goutoi, (together with the Phinnoi, ?Finns) are in 'Scandia', the Scatinavia/Scadinavia of texts (i) and (iv), though at *ibid* 3.5.8, in his Sarmatia section, he has Γύθωνες, Gythones (again with the Phinnoi), so on the eastern side of the Vistula. Iordanes, c 550 AD, himself part Goth, wrote that they originated in 'Scandza' (*Get* III.16–IV.25: his Latin spelling would reflect his pronunciation), a tradition consistent with the Ptolemy references. Tacitus also has the Got(h)ones at *Ann* 2.62; Γότθοι, *Gotthoi*, became the usual spelling in Greek e.g. Procopios *passim*.

19 There are two subsidiary points here. Pliny's sources need not have been entirely literary; he could have acquired information about the Baltic when serving in Germany, and from other officers who had served there. Secondly, we should consider his political outlook on Germany: for him, it had three parts. There was Germania, the occupied areas west of the Rhine, soon to become actual provinces; there was the land further east up to the Elbe, which had briefly been under Roman control, and which he may have felt might still come within the empire; there was the land beyond the Elbe, where the tribes were still Germanic but now unlikely to become Romanised.

20 Cf F17 n 12 for Pliny calling a foreign tribe by its name, and using *genus* for a clan, a group of tribes.

21 It is clear from both Pliny and Tacitus that the Hermiones lived not on the Ocean, but further south.

should identify them with the Inguaeones or the Goths. As noted on F21, in terms of MS transmission Pytheas could have written either. However, since the Teutoni are those who bought the amber, the Guiones must have lived close by, realistically within no more than a day's sailing. The Teutoni, or at least some of them, like the Cimbri, lived on Jutland, and were within the Inguaeones clan. The Guiones' island, therefore, must be close to the eastern side of Jutland. This strongly tips the balance in favour of the name being a corrupted version of Inguaeones, and not of Gutones or some other spelling for the Goths (n 18).

Appendix 5

THE OST- TRIBE(S)

Note: parts of this Appendix are easier to follow with some knowledge of Greek. However, the general tenor of the arguments should be clear to all readers.

When Pytheas was in Brittany, he encountered a tribe well known from a number of Latin texts as well as Ptolemy. When Caesar (or his officers) met them, he represented their name as Osismi (*BG* 2.34.1, 3.9.10, 7.75.4); Ὀσίσμιοι, *Osismioi*, in Ptolemy *Geog* 2.5.8, 9; sometimes spelled with -ss-, as Ossismici in Mela 3.2.23 and F12; for other texts, see CIL XIII 1/1 490. Pytheas also recorded their name, but it is spelled in several ways in the MSS of our references. Until 1963, it was assumed that they were all the same tribe.[1] However, Lasserre (1963) proposed on palaeographical grounds that the Ostideoi of F25, text (b) below, were a German tribe in or near the Rhine. This raises difficulties: (i) we know of no German tribe with a name similar to 'Ostideoi'; and (ii) Lasserre's proposed identification is in itself not easy.[2] But, as shown on F25, Strabo's language shows that he treats the 'Ostideoi' as the Breton tribe, and the reference to Scythia as a separate point on which to attack Pytheas, and that is assumed here.

We have the following:[3]

(a) F23 (Stephanus copying Herodianus): Ὠστίωνες, *Ōstiōnes*: a people by the western Ocean, whom . . . Pytheas [calls] *Ōstiaioi*: to the left of these

---

1 So Mette (1952) 5 n 4: 'offensichlicht . . . um denselben Namen', 'obviously . . . the same name'. Radt takes the same view in his text of Strabo: (2001) 158, 160. But others follow Lasserre: Roseman (1994) 29 and Bianchetti (1998) 129, 205.

2 For technical reasons (see n 10), Lasserre proposes 'Ostidaioi', but that does not affect this aspect. As to (i), see Note to Appendix 4 §3 for Strabo's knowledge of German tribes. As to (ii), Lasserre refers to the Istuaeones of Pliny *NH* 4.100 and the Istaevones of Tac *Germ* 2 (§§5–6 of the Note). The spelling in Pliny is not certain. His MSS mostly agree that the name ends -aeones or -eones, but the first syllable variously appears as istriao-, istrio-, sthreao-, etc (see Zehnacker and Silberman (2015) 74). Editors adopt some emendation to parallel Tacitus' Ingaevones, where the MS tradition is reasonably firm; I follow Zehnacker and Silberman, who adopt Detlefsen's proposal 'Istuaeones'.

3 For readers with no Greek: the name appears here with different endings depending on whether it is the nominative, accusative, or genitive plural. Where practicable, it is put in the nominative.

DOI: 10.4324/9781003181392-9

are the Cossini [also] called *Ōstiōnes*, whom Pytheas calls *Ōstiaioi* [no MS variations]

(b) (c) and (d) are from Strabo; his best MSS, A B and C, are as per Radt's list, vol 1 (2001) VII-IX.[4]

(b) F25: . . . and as to both the 'Ωστιδέους, *Ōstideous*, and what is beyond the Rhine as far as Scythia he [Pytheas] always tells lies . . . [accusative: so spelled in almost all Strabo's MSS]

(c) F26: and the cape . . . of the 'Ωστιδαμνίων, *Ōstidamniōn* [all MSS], called Cabaeon . . . the areas around the capes and of the 'Ωστιδαμνίων, *Ōstidamniōn* [A, B]/'Ωστιμνίων, *Ōstimniōn* [C] [genitive plurals]

(d) F37: . . . there are the οισίσμιοι, *oisismioi* [A]/οἱ σίσμιοι, *hoi sismioi* [B, C; nominatives], whom Pytheas calls τιμίους, *timious* [accusative] . . .

In terms of palaeography, the problem is to decide what Strabo wrote and why text (a) differs; but our interest is what Pytheas himself wrote, and to recover, if possible, how he heard the name when in Brittany and then spelt it in his book: see pp 3–4. Texts (b) and (c) reflect Pytheas via Eratosthenes; it is unclear whether texts (a) and (d) come directly from a copy of Pytheas' book or via some intermediary. It is convenient to start with text (d). That comes from Strabo's geography of France, and it is probable that he will have written *Oisismioi*, or something very similar: his source will have been Caesar or a writer copying him. It is essentially Caesar's spelling with a Greek ending; the *Oi-* vowel in the first syllable, if original and not from transmission, could just be a quirk of someone's pronunciation.[5] The 'nonsense' form *timious* attributed to Pytheas can easily be cured in transmission terms: the letters which precede it are *ous* (the word οὖς, *whom*),[6] and it is clear that a scribe's eye moved too quickly and lost the *ōs* from the beginning of the name; he was copying from *Ōstimious* (nominative *Ōstimioi*).[7] The corruption may be fairly early, if we attach weight to MS A, as if the text had *hoi sismioi* and *hoi timioi*, literally 'the sismioi . . . the timioi'.

Turning to texts (b) and (c), although all deriving from Pytheas via Eratosthenes, they offer three different spellings. Written as nominative plurals they are: *Ōstideoi* in (b); *Ōstidamnioi* and *Ōstimnioi* in (c). The fact the same tribe has different spellings a few lines apart in (c) shows how easy it was for errors in transmission to occur. As to (a), both forms are 'wrong' in that they have lost the consonant sound in the middle, here showing how the spelling of foreign names

---

4 And Diller (1975): A = Paris. gr. 1397; B = Ath. Vatop. 655; C = Paris. gr. 1373.

5 See also the note on Avienus in the text below.

6 The Greek has a ᶜ mark above the *ous*, indicating that, whether pronounced or not, the word began with 'h'; but that can be ignored in the present context.

7 Lasserre (1963) 109 neatly describes this as 'une omission de type haplographique . . . banale de ΩΣ après ΟΥΣ' (*ŌS after OU*S) (so *ibid* 111).

could circulate in different ways. *Ōstiōnes* is interesting in that it has attracted the -*ōnes* ending, a known but not very common form for ethnic names.[8]

It seems clear that Pytheas spelt the first syllable *Ōst-*; given Caesar's *Os-*, it suggests that the local pronunciation was a sibilant sound such as *osht-* or *ots-*, which Pytheas represented one way and Caesar another.[9] We can be sure that there was a consonant sound in the middle syllable, which the *Ōstiaioi* of text (a) has lost: all our other mentions of the name, including Caesar and Ptolemy, have one or more consonants in the middle.

As to this syllable, Caesar used -*ism-*. If we look just at the *m*, it would favour *Ōstim(n)ioi* in texts (b) and (c) as close to what Pytheas wrote. It is a neat solution, and can be supported by noting that confusion in MS copying between Δ and AM (*D* and *AM*) is 'banale': Lasserre's (1963) 109; text (d), and the note on Avienus below, would suggest *Ōstimioi*. But Caesar's inclusion of an -*s-* should alert us to the possibility that the local pronunciation of this syllable was closer to -*isd-* or -*idz-*, which would favour Pytheas writing *Ōstidaioi*. Some prefer that form on palaeographic grounds.[10] The conventions of palaeography should be balanced against our ignorance of how the locals actually pronounced their name, and therefore what Pytheas heard. The most likely candidates for what Pytheas wrote are *Ōstimioi* and *Ōstidaioi*. With considerable hesitation, and following Radt (nn 1, 10), *Ōstidaioi* has been adopted here, the *Ōstiaioi* of text (a), and Caesar's -*ism-* for the middle syllable, perhaps tilting the balance in its favour. On any view, it was an unfamiliar foreign name, and its spelling liable to corruption at more than one stage of its transmission.

There is a final point: the version *Oestrymnici* in Avienus, c 350 AD. His poem is discussed in Appendix 1. He was probably referring to this Breton tribe, despite it also apparently being located near Gibraltar (Appendix 1 p 124). As noted in that Appendix, it is controversial what his sources were, but the many references to Iberia indicate local sources: see in particular Celestino and López Ruiz (2016) 88–91 noted at p 125. It suggests that in the Cadiz area the name was pronounced *Oi-*, and this offers an explanation for the *Oi-* form in text (d); the -*mn-* in the middle syllable proves only that it was one of the pronunciations already circulating in Avienus' time.

---

8 E.g. *Bouprasiōnes*, Steph Byz sv Βουπράσιον; *Cadmeioi / Cadmeiōnes*, id sv Καδμεία.

9 In Pytheas' time, the Bretons were illiterate, and he could not access a local spelling; cf p 3.

10 So Mette (1952) 5 n 4; Radt (2001). The *Ōstideoi* of the MSS is unlikely to be correct on any view; we should amend to *Ōstidaioi*, as ethnic names in this form always end in -*aioi*.

Appendix 6

THULE AND THE FROZEN SEA

It is clear that Pytheas recorded Thule and what for the moment it is convenient to call the 'frozen sea'. Deciding to what they refer depends on several things: (1) where and how he learnt about them; (2) how accurate his informants were; (3) how well he understood what he was told; (4) the reliability and accuracy of our sources, and how closely they reflect Pytheas' actual words. Relevant texts are listed below (not F38, which just mentions Thule).[1] He was the only person who had been in the north, so references to both island and sea must go back to him; but that does not mean that subsequent writers did not elaborate or misquote. The analysis that follows assumes a voyage as proposed in the Coda, pp 170–7: that he overwintered in the Shetlands, during which time he learnt about Thule; and he found the frozen sea in early spring. If he put in at Thule at that stage, he was there only for a day or so. His only experience of summer in the north was in Brittany and south-west England, and the following year in the Baltic. Even if one proposed a different time scale, there is nothing in our references that suggests that he spent time before and after the summer solstice somewhere close to the arctic circle, experiencing continuous daylight for himself. Other views are briefly noted in §17.

Autopsy

1 There is one text:

(i) F30, Strabo 2.4.1 (Polybios 34.5.5): The thing like a jellyfish he saw for himself; the rest he reports from hearsay.

The 'jellyfish' sea is considered in §§14–16, but 'hearsay' carries implications which should be self-evident but are rarely stated clearly: a distinction between what Pytheas saw himself, and what he was told. As noted on F30, Polybios exaggerated his reportage of Pytheas, but here he appears to be accurate. We therefore

---

1 The name also occurs in a number of late authors; none throw light on what Pytheas wrote. For Dicuil and Procopios, see §13.

    DOI: 10.4324/9781003181392-10

have a two-stage problem of reliability: how well Pytheas' Celtic enabled him to understand his informants;[2] and how accurately they, in turn, understood about Thule. For this see §8 below.

Basic location

2 We depend on Pliny and Strabo:

(ii) F15 Pliny 2.187: the island of Tyle, six days' voyage distant to the north of Britain.

(iii) F19 Pliny 4.104: [the] largest [island], Berrice, from which one sails to Thule.

(iv) F24 Strabo 1.4.2: Thule, which Pytheas says is six days' sail from Britain to the north . . .

(v) F31 Strabo 2.5.8: Thule, the most northerly of the British isles . . .

Texts (ii), (iv) and (v) must be from what he wrote; text (iii) appears to go back to him, as discussed in §10. We should first summarise the relevant astronomy.

Astronomical information

3 Pytheas' astronomy for Thule (Appendix 2 §8) goes no further than to place it in the far north. He said that the summer tropic and arctic circle coincided at Thule, and this enabled Eratosthenes to locate it on the arctic circle in our sense (and, in turn, Cleomedes, F2, and Geminos, n 7, to draw on Eratosthenes for their calculations of daylight: see on F2). Also, according to Pliny F15, there are six months of daylight there. If Pytheas wrote that there was a month of daylight, we have lost it. If he visited Thule (§16), it was not in summer, so he would not have experienced it. If, therefore, he wrote that there was six months' daylight, either that is what his informants said, or how he understood what they were saying; and that ties up with what he understood about Thule in general, as next considered.

Thule

4 There are, in theory, four candidates for Thule: the Shetlands, the Faroes, Iceland, and Norway. The first two would be possible if we ignore it being on the arctic circle. Even on that basis, we may eliminate the Shetlands, whatever later writers thought. The six days' distance clearly means continuous 24 hour sailings, and text (iii) shows that his starting point was Berrice, there called an island. You

---

2 The substance here applies if we think that his crew included a Celtic speaker, as has been proposed (p 7 n 9).

would not need six days to sail from John o'Groats to the Shetlands, or even to the Faroes, which are about 275 miles (440 km) as the crow flies from John o'Groats and somewhat less from the Shetlands. The Faroes may also be ruled out because they were uninhabited until the fifth century AD. Berrice is either in the Shetlands, or at the Norwegian mainland opposite: see §11. The realistic candidates, six days away, are Iceland and Norway; we know that Norway is part of the European landmass, but Pytheas would think that any land in the north was an island in the Ocean; that is so whether or not he visited Thule. In view of text (i), we may assume that Pytheas was told about Thule, in practice while overwintering in the Shetlands. What did his informants know or believe, and how did they come by this knowledge?

5 Thule has long been identified with Iceland;[3] but this ignores the realities. Firstly, Iceland was uninhabited until the 8th century AD, when Irish monks settled there, followed by the Vikings; traditionally in 874 AD, but probably from around 800 AD.[4] If the Northern Islanders knew about Iceland, it would mean that they had once sailed into the unknown for many days, well provisioned, with no reason to think that there was any land anywhere to their north; that they happened to sail in the right direction, and in due course happened to come across it; that they did it in high summer (in fairness, a reasonable time to go exploring), and then actually stayed there and were able to support themselves for close to a month, in order to learn that it enjoyed a long period of continuous daylight; and were able to provision themselves on Iceland for their return journey. It is no answer to this scenario that they could have stopped on the unoccupied Faroes *en route*. There are too many improbables to make this an attractive possibility, even taking into account the ability of a skilled mariner to judge where land may be from wind, clouds, sea-swell, and birds.[5] There is also the problem of the type of boat. The Shetlands were essentially treeless, and their boats were withies and skin, i.e coracles. Perfectly sea-worthy in calm waters, they are of limited size (including for food and water), and potentially dangerous in choppy seas. An expedition into the unknown would be very lucky to make it there and back. But they could have acquired a sturdier timber boat from the Scottish mainland or Norway, and that point is not relied on here.

---

3 E.g. Roller (2005) 78–87. A good deal of his text covers attempts to identify Thule, not where Pytheas was and what he wrote; and his assertion at 86 that the diet of F38 relates to the then uninhabited Iceland cannot be right: see on that fragment.

4 The earlier date is evidenced by dating the timber found in the recent excavations at Stöð. Roman (and middle eastern) coins were found, and Roman coins have been found previously. These do not evidence a Roman adventurer, in any case long after Pytheas, but the later ubiquity of coins as tokens and bullion.

5 Roseman (1994) 107 notes that at high latitudes, conditions occasionally cause light waves to curve and enable Iceland to be seen from the Faroes. That would be a valid point here if we assume that the Shetlanders set off into the unknown, discovered the Faroes, and were fortunate that these rare circumstances happened just at that time.

6 In contrast, there are fewer difficulties about Norway. The coastal regions had been settled since at least the Mesolithic: see Østmo (2020) 23–36. His map of flint dagger finds, fig 1.8, p 27, show settlements as far north as the Lofoten islands even at that date, as well as very many finds in the Bergen area; for the Bronze Age he says 'the limit appears be just south of the arctic circle in Helgeland (36)', i.e. up to the Saltfjellet National Park, which is on the arctic circle. There may have been a population decline in the earlier Iron Age, as archaeology becomes more scanty even allowing for possible changes in burial practice; but occupation continued, even if specific sites are hard to identify; generally Østmo, *op cit*, 36–9.

7 What cannot be doubted is that those living further south, and in particular in the Bergen area, including the offshore islands, would know by direct contact or repute that further north there were or had been settlements with long periods of daylight in high summer, about six days' sail distant. We must not be pedantic in seeking a village or farm straddling the arctic circle; nor are we concerned with the calculations of daylight later astronomers offered, or which others believed (cf on F2 and Appendix 2 §8). We are concerned with the actual experience of those living in the area. There are only two places that are literally on the arctic circle: Grimsey island, some 25 miles, 40 km, north of Iceland, and the area of Sand-nessjøen, Norway. But for that sort of distance north or south of the arctic circle you are still in places where there is more or less a month's daylight; even further south, say Huseby (Trøndelag province) in Norway, or Akureyri in Iceland, about 60 miles, 100 km, south of the arctic circle, there is almost a fortnight of continuous daylight, and longer if you include civil twilight.[6]

8 Common sense suggests that there would be some contact between the Shet-lands and Norway long before the Vikings, although there is no archaeological evidence to support that.[7] But the Norwegians had a long history of boat building, substantial enough to enable their settlements to keep in contact with each other: it is well covered in the cited passages from Østmo, above; and they are the obvious source for Shetlanders learning about a distant northern place with continuous daylight. Their contacts would be with a settlement or settlements in the Bergen area, the nearest to the Shetlands.[8] We must allow for the difficulties of accurate communication between the Norwegians and the Shetlanders, and in turn from them to Pytheas, though I assume he had some Celtic (p 7). But he would understand that there was a place six days' sail distant to a place with continuous daylight in high summer, a detail further considered below. Factually, that is

6 Thus Akureyri hosts a midnight golf tournament; there is a bright and short twilight at midnight at the solstice.
7 Private communication from Professor Nils Anfiset, University of Bergen.
8 Not Bergen itself, traditionally founded in 1070 AD, with no archaeology to suggest an earlier settlement at that actual spot.

consistent with reality: the Bergen area is a little over 500 miles from the Mosjøen area. What he said about daylight there is discussed in Appendix 2 §8.

9 The name 'Thule' could help locate it: it is a Germanic name. It is certain that the early settlers in the southern half of Norway were Germanic speakers: see the evidence of old place names in Østmo, *op cit*, 18 n 13 (a long footnote). As to Thule, it recalls *Þulr*, which, at any rate in Old Norse in literate times, referred to a man of status.[9] Grønvik, n 9, argues that *Thule* would be the genitive plural of Þulr, so that Pytheas was recording a place ruled by a group of religious men. He overlooks the point that Pytheas needed to put an 'ē' at the end of Þulr in order to give it a Greek ending, not to record an Old Norse genitive plural; but his basic point may be valid. It is developed by Østmo, *op cit*, 10 with n 5: if later events, the struggles for power in Norway which led eventually to a single ruler, are a guide, there would be local chieftains up and down the coast, and Þulr could reflect the word for such a local ruler. The Shetlanders who heard the word could mistake the name of the chieftain for the name of his realm. At all events, Pytheas' 'Thule' appears to reflect a Germanic name, and supports its location in Norway.

Three other names

10 We now need to consider three other names:

(vi) F19 Pliny 4.104: other islands [around Britain] . . . the Bergi, and, largest of all, Berrice, from which one sails to Thule.

(vii) F13 Mela 3.57: Thule . . . is situated near the coast of the Belcae.

Pytheas is not named in either text, although anything that refers to Thule must go back to him, and for Pliny Bergi were British islands, which again suggests Pytheas. None of these names are known from other texts. The Belcae, text (vii), are problematic for several reasons. At 3.36, when dealing with Scythia, Mela said that it is the name for most Scythian tribes; so he is saying here that Thule is near Scythia. We can explain this in the light of his view of northern Europe, as noted on F13 and suggested on map 2, p 51. But the name is otherwise unknown, as either an individual Scythian tribe or Scythian tribes collectively.[10] It could very well have had the same origin as Pliny's Bergi, text (vi); -l-/-r- and -c-/-g- are easily interchangeable in oral transmission; and we could propose a solution on the following lines. The Shetlanders' Norwegian contacts came from the offshore islands in

9 Grønvik (2006) notes *Þulr* in the Icelandic poem Hávamál, with the connotation of magician, sage, religious leader, and in *Beowulf* 1165, 1456, where English editions offer 'spokesman' and 'courtier'.
10 Collectively, the inhabitants are always Scythian. Greeks knew of many Scythian tribes, though largely clustered around the Black Sea from the Danube northwards and eastwards, e.g. in Herodotos book 4 and Strabo 7.4.2, 5, 11.8.2; as a generic name, 'most are generally called Scythians', Strabo 11.8.2. No name for an individual tribe is anywhere close to 'Belcae'.

the Bergen area (not Bergen: n 8), and Pytheas understood Βέργοι, *Bergoi*, as the name. These islands are in fact hilly (Berg = hill). He would also treat them as other British islands, but not as distant as the last one, Thule.[11] In transmission, the details got confused; for some, the Bergi lost their connection with Thule; as 'Belcae', Mela (or his source) thought it was a tribal name in Scythia, and so placed Thule further east than Germany and opposite Scythia. Any other solution is equally speculative.[12]

11 Berrice, with its Greek ending, also ought to come from Pytheas; it is probably the name of the Shetland island where he overwintered, though it is possible that it was the name of the settlement on the Bergi; from either he would think that Thule was six days away.

12 As a further pointer towards Norway, Pytheas probably treated Thule as inhabited; at least, that is how others understood him. If he visited it as well as being told about it, a possibility envisaged in §16, he would know it. Where he names other places, it is implicit that they are inhabited, and he would not see the need to treat Thule differently. His most northerly British island, text (v), implies habitation. If he had believed it to be uninhabited, he would have said so. Pliny, texts (ii) and (iii), merely noted it as a place; but Eratosthenes understood that it was inhabited, placing it at the limit of the *oikoumenē*, the inhabited world (F25); Strabo, F31, and F38, where he denied its existence as being too cold to be habitable, not because it was uninhabited. As there pointed out, we cannot infer habitation from the British diet, F38; the diet is not that on Thule, but mainland Britain.[13] Cleomedes' 'where they say . . . Pytheas . . . was' (F2) also assumes habitation.

13 We must take care to draw a distinction between deciding where Thule was from the texts that are close enough to Pytheas to enable us to draw sensible inferences, and what later armchair writers thought. Virgil's mention of Thule, noted on p 1, shows that the existence of Thule soon circulated without any specific location. The identification of it with Fair Isle or one of the Shetlands, as noted on F19, was wrong, but understandable by men who knew only their Virgil.[14] Two late authors illuminate this point perfectly. The Irish monk Dicuil (c 770–c 835), writing when Iceland had recently been settled by Irish monks, knew that it was previously uninhabited; his identification of it with Thule was no more than a

---

11 Pliny's 'Bergi' are neutral as to a Greek or Roman origin; Greek nominative plurals in *-oi* are written -i in Latin, e.g. e.g. Sami, Chii, Lesbi, Pliny *NH* 1.5.

12 Bianchetti (1998) 174 proposes that the Bergi 'can be located in the region of Bergen'. Although plural, Bergi has been proposed for one of the Orkney islands: see Zehnacker and Silberman (2015) 336. In the fifth century AD, the half-Goth Iordanes referred to a tribe 'Bergio' as one of many inhabiting the 'island of Scandzia' (*De Orig Act Get* 3.16–24 at 22). But Iordanes was writing when Scandinavia as a whole was much better known, and his Bergio could just be the 'Hill-tribe'.

13 McPhail (2014) 251–4 argues that Thule was in Norway because it was inhabited, much on the earlier lines, though at 253 he proposes that the diet is that on Thule.

14 See the references to Orosius, Rutilius, and Antonius Diogenes, on F19 with nn 6–7.

guess (*Liber de mensura orbis* 7.7–8); Adam of Bremen in the 11th century definitely said that it was Iceland.[15] On the other hand, Procopios, a Byzantine historian (c 500–565) spoke of Thule as a large northern island with many Germanic kingdoms, the context indicating Norway, with 40 days of continuous daylight in high summer; that it was an island shows that it was not yet understood that the landmass further east closed off the Baltic.[16] Neither throw light on Pytheas' actual text.

Frozen sea

14 We now turn to the 'frozen sea': there are four texts to consider.

(viii) F19 Pliny 4.104: From Thule by one day's voyage is the congealed sea, which some call Cronian.

(ix) F17 Pliny 4.95: . . . the Ocean in the north [the Baltic]: Hecataios calls it Amalcius . . . because the name means 'frozen' in the local language. Philemon says that it is called Morimarusa . . . that is Dead Sea, from there as far as the Rusbean promontory, and then beyond that Cronian.

(x) F24 Strabo 1.4.2: Thule, which Pytheas says . . . is near the frozen sea.

(xi) F30 Strabo 2.4.1 (Polybios 34.5.3–5): He also included his story about Thule and those regions in which there was neither land strictly speaking nor sea nor air, but a sort of mixture of these things like a jellyfish, in which he says that the earth and the sea and everything oscillate as if they were bound together, on which you could neither walk nor sail.

For our purposes, we can summarise the concept of a 'frozen sea' as either icebergs and continuous stretches of frozen sea, or ice floes. The 'Cronian' adjective of texts (viii) and (ix), the implications of which are noted on F19, are an addition to or comment on what Pytheas wrote, and suit either kind. In practice, however, it is sensible to think of ice floes; icebergs or fragments of them would probably be too far north to fit the one day's sail from Thule description. Ice floes can take several forms, and there are a fair number of videos available on the web. But

---

15 For Dicuil see Tierney and Bieler (1967) 74–5 (text and translation); 115 (commentary); Roseman (1994) 157–8 for Adam of Bremen. Dicuil, like Martianus Capella, is no more than a repetition of earlier texts. The cited commentary on Dicuil refers to Isidore of Seville's *Origins*, who imaginatively proposed that 'Thule' was so called because the sun reached the summer solstice there; but, as Tierney notes, it is hard to understand; only the -l- in the Greek for 'sun', ἥλιος, *hēlios*, is in Thule. He would know that it could not be the Shetlands; they were occupied and in the diocese of Bergen.

16 Procopios *De Bellis* 6.14–15; 8.20, 25. He is here relating the history of the Herculi, a Germanic tribe who originated in the Black Sea area, and started to come to the notice of the classical world in the third century AD. One can speculate that (1) he had read a copy of Pytheas, and (2) the full text about Thule enabled him, in the light of later knowledge of the Norwegian landmass, to use it as the name of the place with the kingdoms.

one form in particular should be considered: pancake ice. Here, YouTube is particularly helpful, because the images readily suggest that if this is what Pytheas encountered, and he wanted words to describe what, to his Mediterranean readers, was a totally otherworldly phenomenon, the 'jellyfish' description preserved in text (xi) was a good choice.[17] Absent from these videos, taken in clear weather, is the mist or fog which it is obvious that Pytheas encountered, hence 'nor air'. Greek did not have a dedicated word for 'freeze'; they used a word whose primary meaning was 'fix' or 'harden' or 'congeal', πήγνυμι, *pēgnumi*, or a compound of it.[18] Strabo has the same word in text (x), copied from Eratosthenes; Pliny correctly translated this as 'concretum' (F19; the same word in the Varro passage noted on p 3). The 'jellyfish' detail was omitted by Eratosthenes, as an unnecessary addition to what he needed to say, but it has, fortunately, been preserved by Polybios in text (xi).[19]

15 These texts have attracted considerable literary analysis. 'Land, sea, air' are arguably three of the ingredients regarded as the basic constituents of matter: earth, water, air, fire. The verb translated 'oscillate', αἰρωεῖσθαι, *airōeisthai*, is used by Plato in Phaido 111e-112a, where the speaker theorises imaginatively that all the sea and water on earth flows in and out of numerous underground chasms. The phrase translated 'bound together' includes the word δεσμός, *desmos*, meaning 'fetter', a word not associated with philosophy but found a couple of pages before F27 (2.3.5), where Strabo quotes a line of poetry (author unknown) that says that the Ocean flows round the world with no *desmos* to prevent it. But Pytheas was not a philosopher: he was a practical mariner who needed to describe what he saw; though if one is looking for literary comparisons, we could note that the word for 'air', ἀήρ, *aēr*, in Homer means 'mist' or 'haze'. The 'jellyfish' aspect is discussed on F30, and one only needs to look at the videos mentioned above to see what an apt description of this novel phenomenon Pytheas devised for his Mediterranean readership.[20] It cannot be an ingenious invention.

---

17 Of the several available, watch?v=nt690bQqVc includes a useful spoken explanation of their formation, namely from a combination of cold temperatures and waves; it also offers shots of the many shapes possible. See also watch?v=SzrMrd6aSqA. See (e.g.) earthobservatory.nasa.gov/features/Sealce, and nsidc.org/cryosphere/seaice/index.html for ice floes in general.

18 See Chantraine (1968–77) 894; Frisk (1960–72) 525. Herodotos 4.28.1 says that the sea round the Crimea freezes, using *pēgnumi*, but uses κρύσταλλος, *crustallos*, for the resulting ice strong enough to bear wagons as far as Sindica (modern Anapa, Russia: generally Asheri *et al* (2007) *ad loc*; Strabo 2.1.16, 7.3.18; Mela 1.115). When he goes on to refer to a heavy snow-fall, he sensibly adds 'for those who have seen it'. Literary texts, usually written in warm climates, may be a poor guide to how those in the mountains of Arcadia and Thessaly referred to ice: *pagos*, perhaps, the modern word.

19 In total contrast, we find older statements such as Thomson (1948) 148: 'Pytheas can hardly have seen the frozen sea itself; indeed, some think that he need not have gone beyond the Shetlands and interpret the rest as hearsay'.

20 Cunliffe (2001) 129 is correct: Pytheas wanted a straightforward description of the reality. For the literary arguments, see Roseman (1994) 127–31; at n 15 she argues Pytheas' statements about the North Atlantic could not be reconciled with Aristotle's views of the Mediterranean weather in *Meteor*

16 Pytheas would not encounter ice floes in the summer, and it makes sense to think that he came across them in early spring, when he first left the Northern Isles where he had overwintered, and this is assumed in the timetable proposed in the Coda, pp 174–5. It is not, however, clear why he was there. He was seven days' sail from his winter base in the Shetlands; what was he hoping to achieve? At the practical level, how much food and water was he carrying, and where did he think he could get fresh supplies? There is no implication in our references that he put in somewhere in Norway when three or four days out to replenish his food and water. If he had done so, he would think it was another island, and report it. There at least four possible answers. One is that he was told that the sea might freeze in the far north, something the Shetlanders understood from their Norwegian contacts, and he thought that this was worth seeing and reporting. Another is that he wanted to visit Thule, for him the most remote island in this part of the world, but weather and perhaps an inadequate understanding of his direction put him off course. A third is that he aimed to sail over the top of the world, and eventually reach somewhere in the unknown Scythia, incidentally proving that astronomers were correct in stating that the earth was a sphere. A fourth is that he did reach Thule, and that it was there that he learnt that another day or so would take him to the frozen sea; an attractive solution, and explains why Pliny, text (viii), reported it as one day from Thule, not seven days from some part of Britain. He could even, perhaps, have put in at the Bergi after leaving the Shetlands (cf §11–§12); his natural curiosity would encourage him to visit both there and Thule. It solves the provisions aspect, and does not involve completely rejecting Polybios' hearsay statement: he heard about it in winter, and went there in early spring.

Summary

17 There are numerous variant explanations for Thule, the names in texts (vi) and (vii), and the frozen sea, some of which have been noted above; some are based on solid reasoning, e.g. Nansen (1911) 58–68, who argues for Thule as Norway, on the assumption that Pytheas went there; by comparison suggestions noted by Roseman (1994) 94 that the Bergi could be the Shetlands or the Faroes do not seem to have a clear basis.[21] What is proposed above attempts to reconcile the conflicting sources with science and the realities of what is actually there, as well as of sailing in the North Atlantic in Pytheas' time.

346b 21–347a 12, 354b (not 345b) 24–33, and 361b 36–362a 31, and helps explain why so little of Pytheas' text survives. Roller (2018) 95 treats the description as a metaphysical concept: the environment of Thule demonstrated 'a primitive stage in the creation of the earth, where the component parts had not yet been separated, but remained bonded together'.

21 See the discussions and references in Bianchetti (1998) 164–7; in addition to the passages just noted, Roseman (1994) 125–30; 132–3; Zehnacker and Silberman (2015) 336–7. Among general discussions, Thomson (1948) 147–51; Carpenter (1973) 173–84, who argues that Pytheas got no further than the Shetlands; Cunliffe (2001) 116–33.

Appendix 7

STRABO'S VIEW OF WESTERN EUROPE

Many of our Strabo references refer to distances in western Europe. Irrespective of their source and accuracy, they must be understood on the basis of his views of how Iberia, France, Britain, and Ireland sat in relation to each other. It is difficult, if not impossible, to reconcile all the details, and therefore to suggest how Strabo envisaged it. However, maps 3a and 3b indicate how we can make sense of it.

Iberia, he says, is shaped like an ox-hide. The two hind legs are Cape Nerion, Cabo Touriñán, at the north-west, and the Sacred Cape, Cape St Vincent, at the south-west. The Pyrenees form its neck, the eastern side, half the length of the western side (3,000 and 6,000 stades; 340 and 680 miles, 545 and 1,090 km).

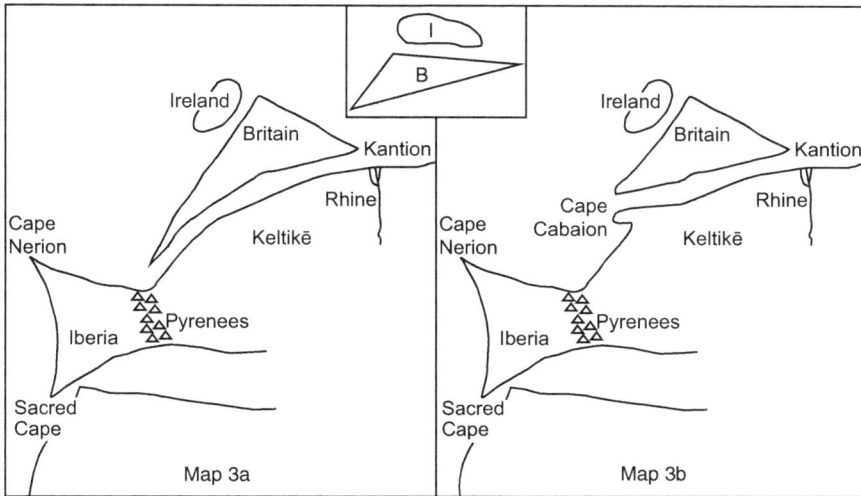

Map 3a Sketch maps of northern Europe as Strabo apparently presents it; *map 3b* possible alternative layout allowing for Cape Cabaion; *inset* alternative location for Ireland, if thought of as parallel to the 'east' coast of Britain.

Source: Drawn by the author.

DOI: 10.4324/9781003181392-11

The Pyrenees run north and south, and separate Iberia from Keltikē. He does not explain how the south coast comes close to Africa at the Pillars of Heracles, Gibraltar, though he does say that it faces both the Ocean and the Mediterranean (2.5.27, 3.1.3; cf F34). Even so, in itself it is a reasonable picture of Iberia. But when we see how he places Britain in relation both to Keltikē and Iberia, there are problems, both of contour and distance.

At 4.5.1, at the start of his description of Britain, he says that it is triangular, with its longest side parallel to Keltikē, each being 4,300 or 4,400 stades (490, 500 miles; 785, 800 km). Putting the longest side opposite France looks wrong to us. He knew of Pytheas' dimensions for the three sides, to judge from his wrongly citing Pytheas' 20,000 stades for the south coast, the one opposite France, in F26; presumably Eratosthenes and Hipparchos had used Pytheas' figures for Britain. As we see from F4, 20,000 stades was Pytheas' dimension for the west coast, up to John o'Groats, and it was also its longest side. It is as if Strabo's antipathy to Pytheas made him want to project a picture of Britain that was different, hence his making the coast opposite Keltikē the longest side; cf his blanket condemnation of Pytheas' northern geography, F26. He either did not know, or ignored, Caesar's dimensions noted on F4, which, like Pytheas, made the south coast the shortest side.

Further, Strabo describes the Keltikē coast as running from the mouths of the Rhine to the northern end of the Pyrenees by Aquitania. It is pictured as a straight line or gentle curve from the Aquitanian side of the Pyrenees to the Rhine, a belief which Roller (2018) 200 points out persisted to 200 AD. So the longest British coast 'lies alongside' the Keltikē coast, from Kantion, Kent, to its western end, which is 'opposite Aquitania and the Pyrenees'.[1] As proposed on F25, the view that Kent was opposite the Rhine probably developed from the belief that if Britain was opposite Keltikē, and Keltikē extended to the Rhine, then Kent must be opposite the Rhine. Strabo's description of the Keltikē coast partly corresponds to Caesar's description, *BG* 5.13; but, as noted above, the latter is clear that it is the shortest side which runs west from Kent. But Caesar has Land's End pointing south, with the west coast facing Spain, apparently making Britain a sort of isosceles triangle leaning over to the left; his south coast is only some 500 miles or 800 km long. While Strabo's dimension of 4,300 or 4,400 stades is reasonable for the south coast of Britain, it is hopelessly wrong for the European coast from the Rhine to the Pyrenees. It is a reasonable approximation for the northern coast, from Brittany to the Rhine, and equally from there south to the Pyrenees, but as a total distance it cannot be justified even as a mariner's estimate.[2]

That leads to question how Strabo fits Brittany in to this picture. He knows that it is occupied by the Ost- tribe, and that it ends with Cape Cabaion: F26. More

---

1 The Greek word is παραβάλλω, *paraballō*, commonly translated 'lies parallel to'.
2 Even allowing for his mistake at 4.3.3, that it is a short distance from the Seine to the Rhine.

specifically in F37 he says that the headland protrudes 'quite far': see *ad loc*. At F25 he puts the British south coast at the slightly higher figure of 5,000 stades (570 miles, 910 km), again the same length as that of Keltikē opposite, but with no mention of the Pyrenees or Aquitania.[3] The 5,000 stades figure shows that, as elsewhere, various distances circulated (cf p xvii), and in the present context the difference does not matter. What is important is that these details suggest a picture on the lines of map 3b. This retains the south coast of Britain as the longest side, but with a realistic length for the European coast opposite.

On the above approach, he used different sources to Mela, who basically got Britain right (F12).[4] We could explain the 'opposite Aquitania' reference in 4.5.1 as he or his sources picking up mariners' talk of sailing west down the English Channel and south to the northern coast of Spain or Aquitania. They would say that once clear of Land's End and Cape Cabaion, the run to Oiasso (Irún) or Lapurdum (Bayonne, Aquitania) was more or less direct. Either is about 500 miles, 800 km, and could easily have been expressed as 4,300 or 4,400 stades. A modern map, as map 4a, p 178, shows that it would be easy for them to treat the western tip of Britain as opposite Spain in that sense. Strabo or his sources built on that, and led to Strabo expressing himself as noted above.

Strabo's location of Ireland is uncertain. At F25 it is north of Britain, more precisely less than 4,000 stades north of the middle of Britain (−455 miles, −730 km), so F31; at 4.5.4, within his geography of Ireland, he says that it is parallel to and north of Britain; but he does say whether parallel to the south coast or one of the others. Map 3 offers alternative possibilities.

Turning to distances, we have the following ([*H*] = from Hipparchos; [*S*] Strabo's own figure; [*?*] source uncertain):

(a) Marseille is on the same latitude as Byzantium [*H*] (F25, F27; 2.4.3; with reservations, F31);

(b) the latitude through the mouth of the Dnieper is 3,700 stades (420 miles, 670 km) north of the Byzantium latitude [*H*] (F27, 2.1.16); 3,800 stades (430 miles, 690 km) [also apparently *H*] (F31);

(c) the Dnieper latitude runs through Britain [*H*] (F25); through Keltikē [*?*] (F27, F28);

(d) the Dnieper latitude runs through Britain [*S*] (F31);

(e) from Marseille to the middle of Britain is no more than 5,000 stades (570 miles, 910 km) [*?*] (F25);

---

3 The Greek is παράκειμαι, *parakeimai*, 'lie alongside', with the same connotation as *paraballō*, n 1.
4 All Strabo's MSS have 'longest' [side], and it is difficult to see that the word is a very early error for 'shortest'.

(f) another 4,000 stades (455 miles, 730 km) is the northern limit of the *oikoumenē*, Ireland (presumably its northern coast) [*S*] (F25);

(g) Britain is 2,500 stades (285 miles, 455 km) north of Keltikē [*S*] (F28);

(h) 6,300 stades (715 miles, 1,145 km) north of Marseille is in Britain, not Keltikē as Hipparchos [*S*] (F28).

It is not clear how far a [*?*] statement goes back to Eratosthenes or some later source; or whether an [*S*] figure is Strabo's own calculation or taken from one of his sources. Unlike his predecessors, he was writing when Gaul was under Roman rule, and official figures, certainly road distances and perhaps overall dimensions, would be available. The analysis following suggests that he did not use such information, but sources depending essentially on traders' talk and the like. He did use oral sources if there was nothing in writing (2.5.11).

Given these differing figures, it is impracticable to reproduce them on either map, much less a modern one. He does not offer a distance for the width of Keltikē, e.g. from Marseille to the English Channel, though the above figures suggest that he thought that Keltikē was narrower than it is.[5] He records Polybios' 40,000 stades for the circumference of Britain, F30, and, as noted above, knows Pytheas' individual dimensions for the three coasts. Strabo only offers figures for the south coast; see above. We can, however, build up some sort of picture as to his belief for the area. Taking texts (b), (c) and (d) together, on balance it appears that he thought that the Dnieper latitude ran through Britain, not Keltikē. On that basis, he appears to have thought that (1) 3,700 or 3,800 stades north of Marseille was in Britain, text (b), and by implication (2) the English Channel was less, say 3,500 stades, 400 miles, 640 km, from Marseille; (3) from somewhere in Keltikē to somewhere in Britain is 2,500 stades, text (g); from Marseille to the middle of Britain is 5,000 stades, text (e); from Marseille to somewhere else in Britain is 6,300 stades, text (h); (4) 4,000 stades north of the middle of Britain is the limit of the *oikoumenē* (F25; cf F28, F31, 2.1.17). He does not indicate whether the apex of his British triangle is opposite to or south of Ireland. Even so, his accepting Pytheas' British diet, F38, suggests that he regarded most if not all of Britain as within the habitable zone.

---

5 As the crow flies, the distance is typically between 500 and 550 miles, 800 to 880 km.

CODA

1 Our basic material – the references

When one looks at our references to Pytheas in the round, they are a mixed lot, in more than one sense. As we read him in Diodoros, he is twice compressed, firstly by Timaios, and then by Diodoros' epitomising Timaios. Polybios and Strabo were hostile and report him adversely. Strabo criticises Eratosthenes and Hipparchos for believing him, and says that Pytheas lied about areas in the north. What we read of Pytheas in Mela comes via Eratosthenes: Silberman (1988) 282–3. Pliny accepted Pytheas, but suffers from his method of working from notes; details in his sources are lost, and it increases the risk of unfamiliar foreign names being mangled when written in his book. Scientists believed Pytheas, but Eratosthenes and Hipparchos survive only as fragments; Cleomedes (F2) and Geminos (F7) tell us something about his astronomy but little about his route and his description of it. They are a mixture in another sense: they present him here as a mariner and intrepid explorer, there as a curious and interested enquirer and observer, elsewhere as a geographer, or an astronomer; the references to tin and amber link him to the trade in them which passed through Marseille as well as suggesting one reason for his voyage. But it is hard to flesh out the limited skeleton of what we can have without a fair number of assumptions.

2 Pytheas, the man and his boat

As proposed at p 7, we can infer that Pytheas came from a family who ensured that he had a good education; and if not themselves engaged in trading, they knew those who were. They would help him in acquiring a sturdy boat suitable for long sea voyages (see below); the help would include supplies of wine and perhaps other items which he could use to pay his way *en route* for food and other needs, for instance repairs to his vessel. To that we can add experience of seamanship and navigation. As to Polybios' hostile comment in F30, that Pytheas was a 'poor man', as there noted Polybios had both an aristocrat's disdain of social inferiors and a motive for wanting to attack Pytheas. This exaggerated sarcasm tells us much about Polybios' character, but nothing about Pytheas and his family background.

DOI: 10.4324/9781003181392-12

Given the penteconter tradition of Phocaea and, in turn, Marseille, the reference to 'us', meaning Pytheas and his crew, in F7, and the realities of what he did and where he went, it is reasonably certain that his boat was powered with both oars and sails. Whether it was as large as a penteconter,[1] with the concomitant problem of feeding a large crew for a long period in unknown territory, is a question; it would also be less expensive for those supporting his voyage to provide a smaller boat. A crew of as few as 10 or 12 would be sufficient to keep the boat moving if the winds were adverse or absent, but it must also have been quite sizeable, probably not less than 50–60 feet long and 15–20 feet broad.[2] It is possible that his crew included a Celtic speaker, though he probably had some Celtic himself (p 7 and p 7 n 9). He may have referred to it as his ἄκατος, *akatos*, suggested to have been the Greek equivalent of *actuaria* for a boat propelled with both sails and oars (so Arnaud (2005) 122), or his λέμβος, *lembos*, the word for a boat smaller than a triaconter.

3 His dates

We may put his voyage as between c 330 and c 325, with his book being published a year or two later. This suggests a birth date of c 360–350. There are three factors which support this period. One is the voyage itself; it suggests a youngish man in his 20s or early 30s, but old enough to have good maritime experience. The second is his considerable astronomical abilities, evidenced particularly by his being able to locate Thule where the summer tropic and arctic circle coincided, and the possibility of using *gnomon* readings to establish latitude. This suggests that he was born late enough for his education to include some of the learning of the astronomer Eudoxos of Cnidos, 391/90–c 337: see pp 19–22. We may therefore think of him as a teenager after c 340. On one view of F8, he said that Eudoxos was wrong in relation to the Pole Star, but, as there noted, it is not clear whether we can attribute those words to him; in terms of dating, it is sufficient to infer that he had learnt Eudoxos' advances in astronomy. The third is the date of publication. His book was not known to Aristotle, who retired in 323 and died in 322; but the works in which we would look for a Pytheas reference such as *De Caelo* and *Meteorologica* had been written some time earlier, so that of itself would not preclude Pytheas' book being published by (say) 330. More positively, F30 shows

---

1 Even if 'penteconter' could refer to a boat with fewer than 50 oarsmen: p 5 n 2.

2 Others have naturally thought of it being a penteconter. Carpenter (1973) 152 quotes a former president of the Royal Geographic Society, Markham, who wrote, 'A large Massilian ship was a good sea-boat and well able to make a voyage into the Northern Ocean. She would be from 150–170 feet long, with a depth of 25 or 26 feet and a draught of 10 to 12. Her tonnage would be 400 to 500, so that [it] was larger and more sea-worthy than the crazy little Santa-Maria with which eighteen hundred years later Columbus discovered the New World' Hawkes (1977) 44 also proposed a penteconter (mistakenly referred to as a trireme by Roseman (1994) 148). Sturdy and sea-worthy, certainly, and suitable for the Atlantic; but we may doubt a crew of as many as 40 to 50.

that his book was referred to by Aristotle's pupil Dicaiarchos, born c 375, and F4 and F5 show that it was used by Pytheas' younger contemporary Timaios. The fact that Dicaiarchos knew it suggests that it was available by c 320, in turn indicating the voyage as around 325. One could add here the real possibility that Pytheas' contemporary Hecataios of Abdera's *On the Hyperboreans* was a parody of Pytheas, as proposed by Hawkes (1977) 38–9; see on F17 for a summary of its details.

The likelihood that this range of dates is correct can be supported by other arguments, but equally it is not impaired if they are rejected as ingenious rather than likely. They build on what is recorded about Alexander the Great. He was very interested in the geography of areas beyond where he was; not just routes from the Persian Gulf to India and Arabia, but further west, such as whether the Jaxartes, which in fact flows into the Caspian Sea, was the upper reaches of the Don. He believed that the Ocean surrounded the Europe-Asia-Africa landmass, and a year or so before his death he ordered that the Caspian be explored to see if it had an outlet to the north into the Ocean.[3] As noted on F17 §94, belief that this was so persisted. This has led to the suggestion that Alexander sent Pytheas to investigate the north: Dion (1977) 180–9, sensibly rejected by Bianchetti (1998) 31–3; a slightly more plausible view is that Alexander learnt about the Ocean and the north from Pytheas' book (which would imply that it was published, and a copy reached him, by 325).[4]

A different use of Alexander's plans stems from the evidence that in 324 Alexander said that he would go no further east than his Indian frontier, and gave his general Crateros written instructions to prepare for an invasion of the Mediterranean, starting along the north African coast from Egypt towards Carthage. They included building 1,000 ships larger than triremes in the eastern Mediterranean and a road along the north African coast. It is set on in detail in Diodoros 18.4.1–6; more briefly in Arrian 7.1.1–4, Curtius 10.1.17–19. The reliability of this evidence is disputed, but Arrian, having said that 'some say' that Alexander had these plans, goes on to say that they were consistent with his character: 'none of his plans were small and petty' (Brunt's Loeb translation). Badian (1967) showed that the plans are genuine, in terms similar to Hammond (1980) 300–6.[5] At the very least, they

---

3 Arr *An* 7.16.1–3 (building ships on the Caspian for the purpose); generally Hammond (1986) 629, 636. Plut *Alex* 68 attributes instructions to sail round Africa, returning via Gibraltar; at least that shows that stories of his wider ambitions circulated.

4 Dion's chapter IV (175–222) is entitled 'Alexandre le Grand et Pythéas', and proposes a route and other details. See Bianchetti (1998) 34–6, citing, among other authorities, Seneca Q.N. 5.18: 'Alexander . . . quaeretque quid sit ultra magnum mare' ('and Alexander enquired what was beyond the great sea'). Roller (2006) 58–60 explores the possible Alexander connections at some length, and concludes 'how much [Pytheas' voyage] reflects a Massilian reaction to Alexander cannot be known' (60).

5 Perhaps by an oversight, in *id* (1986) 639 n 1, 643 n 1 Hammond preserves the doubts of his first edition.

indicate that it was widely believed that Alexander planned a *Drang nach Westen*, and it is inferred that their gist was known when various embassies, including one from Carthage, came to Alexander earlier that year, if not sooner.[6]

Hawkes (1977) 42–4 proposed an actual treaty in 326 between Carthage and Marseille on the basis that Carthage wanted allies against Alexander's plans, and dates the voyage to 325; but (a) in 327–6 Alexander was fully occupied in pushing his conquests ever further eastwards. He would only have started talking of plans to move west once he had decided where his eastern frontier was; (b) assuming he had spoken of long-term plans earlier than 324, and rumours of them eventually reached the Mediterranean, it is doubtful if they were so definite that it caused Carthage to make such a treaty. One could argue for a treaty in 324–3, after the Carthaginian embassy returned from Babylon, which the Carthaginians could feel honour-bound to abide by after learning of Alexander's death. Alternatively, a treaty, or at least some understanding, could be argued on a different basis: Marseille had for some time been on friendly terms with Rome, and Carthage could see that if she also had friendly relations with Marseille, it could indirectly offer some protection against potential Roman expansion.[7] Whatever the merits of any of these proposals concerning Alexander, they do not detract from the basic range of dates proposed.[8]

We should, however, note that other dates are proposed. Roseman (1994) 155 puts the voyage around 350 or earlier, with the publication of the book delayed to c 320; she stresses that Pytheas collected data, and argues that theories to explain data started only a generation or so later; she does not deal with Pytheas' astronomical education, though at 118, dealing with our F8, she assumes that Pytheas was correcting Eudoxos. Others, with the firm belief in the complete closure of Gibraltar, have later dates: Cary and Warmington (1929) 33 proposed 310–6, when Carthage was preoccupied with defending the city against Syracuse; 'plausible' per Roller (2006) 66;[9] Carpenter (1973) 147–50 argued for 240–38, on the basis that this was after Carthage conceded the sea west of Sicily to Rome; he assumed that Marseille's treaty with Rome was already in place.[10]

---

6 The Carthaginian embassy was one of many that went to Babylon in 324: Arrian *An* 7.15.4–6; Diodoros 17.113.1.
7 Strabo 4.1.5 records friendship between Marseille and Rome. Diod 14.93.4 says that the Romans' dedication of the spoils of Veii (396) were deposited in the Massiliot treasury at Delphi (though Livy 5.25 just says deposited at Delphi); Justin 43.5.6–10 says that Marseille sent gold and silver to Rome to replace the cost of buying off the Gauls when they occupied Rome in 390, adding that it was then that a treaty (*foedus*) was made.
8 Most imaginatively is the argument that Pytheas' book was published before Alexander captured Tyre in 322, with details in the late fiction of Antonios Diogenes noted on F19 (see n 7 there).
9 Who adds, however, 'if one believes that Pytheas passed through [Gibraltar], which seems unlikely'.
10 He argued that the reference to Dicaiarchos not believing Pytheas in F30 stems from an error by Strabo, and comes from Eratosthenes noting 7,000 stades from Marseille to Gibraltar, and disagreeing with Dicaiarchos' 7,000 stades from the Straits of Messina to Gibraltar, as recorded in Strabo 2.4.4.

4 Overview of his voyage

This overview is designed to deal with three connected aspects of Pytheas' voyage: what distances he needed to cover, how long it took him, and how he calculated the dimensions of the coasts of Britain. There is little, if anything, in our references that helps on these points. The stage by stage breakdown and timings in section 5 proposes two years: Marseille to the Shetlands in year 1, where he overwintered and serviced his boat, and year 2, Shetlands to Kent and the French coast, on to the Baltic, and home. He was a practical mariner as much as an astronomer, and he did not have infinite resources; except between Gibraltar and Marseille, and possibly along the Atlantic coast of France, he would not have had *xenos* contacts in local ports who could help.[11] There would be a limit to what he could carry to help pay his way (cf p 7). No doubt in some places the locals were generous to passing travellers, but over the expedition as a whole he would need to make gifts or payments to maintain himself, his crew, and his boat. In the Baltic, he had reached the limit of his exploration, and his crew, unaccustomed to being away in the winter, would be glad to get home again; they would ensure that they made good time on the return journey, and return home by the end of the second year's sailing season.

Others take a different view. The possible five years which Roller (2018) 68–9 proposes is noted and rejected in Appendix 2 §§11–12. As long as six years has been proposed: see Dicks (1960) 186–7. If it were such a long period, we could have expected Strabo to attack him for the length of his expedition and not just for his 'lies' about the north. When Polybios attacked Pytheas for being too poor to travel, F30, he did so from the standpoint of his own journeys, each of which was within a sailing season; even so, he complained about the expense of travel (F30 n 2). If Pytheas had been away for several years, that would be an extra stick with which Polybios could have beaten him.

Distances for each section of the route, and corresponding timings, are proposed. The distances are ballpark figures on the basis indicated at pp xvii–xviii. As there stated, they must be read as approximations, but they are realistic if a distance of x miles or kilometres is read as 'in the region of' that distance. As to his speeds,

He also argues that Timaios did not know of Pytheas, based on the discrepancies about the six days taking tin to 'Mictis' in F19 and the island names in F21. He overlooked the details in F4 and F5; no one other than Pytheas could have gone into such detail.

11 For those unfamiliar with the *xenos* ('guest-friend') concept, it was common to have ties of hospitality with a family in another port with whom you could stay when travelling and who would support you financially and practically, and for whose members you would do the same when they came to your home. It was of especial value to merchants and mariners. In Homer, the ties are between élite families, but Hdt 6.21.1 shows that it was established commercial practice, probably long before his day: Scott (2005) 125–6. As to a *xenos* tie between Greeks and non-Greeks, our literary references are all to élite familes, but a commercial tie between Pytheas' family and a Celtic family in the Bordeaux area is feasible: cf on F5 §4.

it is clear that a well-built ship in good hands and with reasonable weather could be expected to cover some 70 miles, 110 km, a day, and Pytheas probably achieved that on a few stretches. But we should not assume that he was aiming to complete his voyage as quickly as possible, and in terms of assessing his overall progress a slower rate of progress is suggested (pp xvii–xviii). Over the sort of periods with which we are concerned, good days and poor days would average out. Once past Gibraltar, the coasts were unfamiliar, and he would proceed cautiously; from headland to headland where practicable. On his outward journey, it is assumed that he normally sailed only by day. He would not need to find a harbour or beach at night; he would anchor offshore.

He was surprised how much longer it took to go round Iberia compared to the overland journey (F34); that indicates that he was making such progress round Iberia as he could. Unless there was something to detain him: information in Corbilo about Britain, F36, the tin mines in Cornwall, F4, and amber in the Baltic, F21, for instance, it is likely that he would usually want to make reasonable progress. The timetable proposed next shows that it would be easy to reach the Northern Isles, most probably the Shetlands, before the end of the first year's sailing season: the distances proposed total some 5,575 miles, 8,920 km. If he left just after the spring equinox, he would have about 220 days to the end of October, and allowing even 20 days on land for one reason or another he would only need an average of 28 miles, 45 km, per day. Three details support his overwintering in the far north, in practice the Shetlands: it was there that he learnt about Thule (Appendix 6 §§6–8), and his being shown the 'sun's bed', F7. As there noted, that must be in winter; if he was there in summer, he would experience it himself. The third is his encountering ice floes further north; that must be very late or very early in the year, and in practice the latter; it would be the end of the sailing season by the time he reached the Northern Isles. As proposed in Appendix 6 §16, he would learn about a frozen sea while there; when he encountered it, it would at the end of winter or the very beginning of spring.

To complete his voyage, he would need to cover about 2,715 miles, 4,345 km, from the Shetlands south to mainland Europe and then east to the Baltic, and 5,280 miles, 8,450 km, from the Baltic back to Marseille; total (say) 7,995 miles, 12,790 km. It is not feasible to offer an average rate of progress for these two sections; he would make gentle progress on the outward journey; once he was returning home, he would make quicker progress, partly because many of the coasts would be familiar from the outward voyage, and partly because of the incentive for him and his crew to get home. He probably had some overnight sailings when returning where this was practicable.

5 Pytheas' itinerary in detail

His route as proposed here is shown on map 4a, p 178. The distances have been calculated as indicated in §4, and are offered as realistic approximations.

(1) Marseille to Gibraltar

Common sense suggests that he would leave in the spring, and it is proposed that he noted the *gnomon* reading for Marseille at the spring solstice and then set off (Appendix 2 §2; cf on F25). At least some of this stage, down the Mediterranean coasts of France and Spain, would be familiar, if not to Pytheas himself, then to some of his crew; in any case its harbours and any particular perils would be common talk in the maritime community at home. Some of that talk might extend to the stretch past Gibraltar as far as Cadiz: the difficulties of knowing what Massiliot traffic there was through Gibraltar have already been noted (pp 15–16).

This stretch involves some 1,165 miles, 1,865 km, and would need 10 days of continuous day and night sailing with favourable conditions. No doubt he was keen to make good speed here, and he may have done one or two 24 hour days at the start, where the coasts were familiar; but it is prudent to allow him some 17 to 20 days for the whole stretch. If he left Marseille just after the spring equinox, he would pass Gibraltar around the middle of April.

We should note Strabo 2.4.4, who prefers Eratosthenes' 7,000 stades (795 miles, 1,270 km) for Marseille to Gibraltar rather than the more than 9,000 stades (1,020 miles, 1,630 km) in Polybios. Carpenter (1973) 148, as part of his argument for a late date for Pytheas (see above), proposed that Eratosthenes took his figure from Pytheas in order to correct an earlier figure by Dicaiarchos. If Pytheas really put this stretch at 7,000 stades, it would mean his sailing in a direct line away from the coasts, in seven day and night sailings, and knew the shortest practicable route available to modern shipping: it is about 690 nm (795 miles, 1,270 km), coincidentally equivalent to Eratosthenes' 7,000 stades. Polybios gave a detailed breakdown of this distance at 3.39.5–7, totalling 9,600 stades, 1,090 miles, 1,745 km, from Gibraltar to the Rhone, and this is the sort of distance Pytheas would need to cover in practice. As a further objection to Carpenter's point, it does not follow that Pytheas felt the need to record this stretch of the coast in detail.

(2) Round Iberia

F34 and F35 enable us to trace him round Iberia; from Cadiz to Cape St Vincent, and round Iberia to the Pyrenees, where he noted that it took him much longer than going overland. To circumnavigate Iberia, he would need to cover some 1,410 miles, 2,255 km, equivalent to some 15 to 20 days' continuous overnight sailing. But it is not a coast that lends itself to that, except to an experienced mariner who knows it well, and we should assume that he covered it mostly if not entirely by sailing only by day. The prevailing currents are broadly from north to south and would not help on his outward journey. He recorded 5 days from Cadiz to Cape St Vincent (F34). We should allow some 5 or 6 weeks, including occasional rests and short days, and we may put him in the San Sebastián area at the end of June or early July. There is, incidentally, no evidence that he spent time on shore to report on the local diet or way of life. He would be passing the north coast of Spain at the summer solstice.[12] F35 indicates that he said something about the Atlantic coast of Iberia, but other than the quicker overland route of F34, it is lost.

(3) The Atlantic coast of France

We can now follow him up the Atlantic coast of France to the Loire estuary at Corbilo (F36). On this stretch, he would find the locals similar to the Celts at home, and nothing which merited a comment. Some of those bringing tin to Marseille were likely to have come from there, particularly the Bordeaux region: see on F5 §4, and §4 above for the possibility of a *xenos* connection. This stretch is about 380 miles, 610 km, about 6 to 8 days' sailing. At Corbilo, we can certainly infer that he stopped for a few days, learning something about tin from the island which he recorded as *Prettania* (F5); he was told that it was four days' sail distant (F4, F36). We may suggest that he would arrive in Corbilo around mid-July, and leave a week or so later.

(4) Across to Cornwall

What he learnt at Corbilo determined his next direction. Until he learnt that Britain existed, he probably envisaged continuing around the Ocean coast of Keltikē, France, enquiring about the tin islands (pp 6, 8–9). He now knew that he had to visit Britain. The four days' distance assumes mariners familiar with the route. Pytheas would need longer. The whole of the Breton coast up to the English Channel is sinuous; it is full of bays, large and small, and the stretch around Pointe du Raz is dangerous (see on F26 with n 5). A prudent mariner unfamiliar with the route would need to sail close to the shore and proceed with caution (unless we imagine that Pytheas had a local on board to act as a guide). He would almost certainly want to sail only by day. He would need to cover about 400 miles, 640 km, of which about 270 miles, 430 km, was up the Breton coast and 130 miles, 210 km, across the English Channel to the Penzance area. We should allow him six days, arriving in Cornwall around the end of July. As he sailed round Brittany and passed Cape Cabaion, Pointe du Raz or possibly Pointe de St Matthieu, he either visited Ushant or at least heard about it, and noted it. It was three days' sail away, but Strabo does not tell us from where, or whether Pytheas went there or merely heard about it (see on F26). Since the offshore island to which tin was taken is St Michael's Mount (see on F5), he would touch land in the Penzance area. It would now be the end of July or the beginning of August. He was the first literate person to visit Britain and introduce it to the classical world.

---

12 This view will be anathema to those who want him to note a summer solstice reading much further north.

He clearly spent some time in Cornwall. Quite apart from the desirability of giving his crew a rest, he found the locals particularly civilised and hospitable, and noted this (F5). They showed him how tin was extracted, and he gave a detailed account of it; if he hoped to break into their long-established export trade he was disappointed (see on F5). He did, however, learn that the cape at the far end of their territory, Land's End, was called Belerion (F4). We may envisage him not leaving until the middle of August.

(5) From Cornwall (Penzance) to Scotland and the Northern Isles

He now had a choice of directions: north (i.e. up the west coast) or east (along the south coast). It is probable that he chose north, perhaps by chance, perhaps learning about favourable south to north currents in that direction.[13] It also makes it easier to explain the rest of his voyage, and how he was able to recognise that Britain was triangular, and give dimensions for its three sides. As he also reached the Baltic, it makes sense to think that he first sailed north from Cornwall up the west coast, and then down the east coast; from there, he investigated what lay further east in mainland Europe and reached the Baltic; he eventually returned home along the English Channel. That is a realistic itinerary, and if correct we should note one detail: it would mean that his figure for the length of the south coast was not from sailing along it himself, but by assuming that it was more or less the same as the distance along the French coast opposite, as he sailed westwards on his way home; cf on F4.

In due course, he reached the John o'Groats area. Estimating his distance and timing from Cornwall is difficult, for several reasons. One is the sinuous nature of the west coast. Many stretches are reasonably straightforward, and he could make good progress. But how far up the Bristol Channel did he sail before he felt able to cut across to the other side? How long did he take to negotiate Morecambe Bay? When he reached the Girvan-Ayr region, it would take time to find out that the Firth of Clyde has several culs-de-sac; once he reached the Hebrides, he would have to find his way through them. Without a chart, he would take time to negotiate round the Arran-Islay-Mull area.[14] Then there is the statement in F30 indicating that he spent some time on land, away from his boat. Britain was new and therefore mysterious, and the local way of life was worth investigating; apart from F5, there is the diet and the other details in F38. We may infer that he spent less time on shore than Polybios' exaggeration would suggest, but one or two such trips while coming up the west coast are possible. It does not mean that he spent long periods away from the coast, but we should allow for some delays. Overall, the whole coast was unfamiliar, and he almost certainly would

13 See the map showing currents in Dion (1966) 206 (fig 1) or (1977) 192 (fig 14).
14 There are difficult tidal currents that need great care to negotiate (communication from Dr Peter Hogarth, from personal experience).

have covered most if not all in daylight only. The question whether he noticed Ireland is noted below, pp 181–2.

Given the many bays and estuaries, he would need to cover some 1,920 miles, 3,070 km, to John o'Groats, and a further 300 miles, 480 km, threading his way through the Orkneys until he reached the Shetlands, making some count of their number. During this final stage, he encountered the huge tides recorded in F16, typical autumn weather for the area. He would be somewhere in Scotland at the autumn equinox, probably in the Portpatrick area. Leaving Penzance in the middle of August, he would still be able to reach the Shetlands at the end of October and lay up his boat for the winter. If we assume as many as 14 days on land, which is probably more than he actually spent, he would need to average only some 36 miles, 58 km, per day.

That leaves the question as to how he came to estimate the length of the coast at 20,000 stades, 2,270 miles, 3,630 km (F4). As suggested on pp xvii–xviii, his rule of thumb was probably 700 stades (80 miles, 130 km) per day for straightforward journeys in the Mediterranean. But this was an unfamiliar coast, and he probably did not achieve quite as much even on the straightforward stretches of the west coast; as noted above, there were many stretches where he would make slower progress. But if we look at it from his perspective, he could not measure either how many hours he had sailed on a given day (cf Appendix 3), nor how many stades he had achieved. The most he could do was to assess each day's progress in the light of his Mediterranean experience and then express the total as a round figure. We can only guess how far he worked out the distance of the west coast of Scotland in the light of his progress around the various islands there, and whether he accepted that the distance should allow for the time he had had to spend negotiating them. A further detail is whether he kept a log or relied on his memory to arrive at his figure: this is discussed below. His figure is about 10% higher than the actual distances proposed on F4, and his estimate cannot be said to be unreasonable.

(6) Overwintering in the Shetlands

The balance of probability is that he overwintered in the Shetlands, rather than in the Orkneys, because it is more likely that it was the locals in Shetland who knew about Thule from contacts with the Bergen area (Appendix 6 §8). Wherever he was, with the end of sailing season it was time to rest his crew, and see to the maintenance and repair of his boat. He would certainly find out how the locals coped with the winter. He was shown where the sun rose and set in high summer; this is clear from F7. He will have heard talk about a frozen sea: Appendix 6 §16.

(7) Departure from the Shetlands

We may assume that he left the Shetlands twice. At the end of winter, he sailed north for seven days until his way was blocked by ice floes; he may have put in at

Thule *en route*: see Appendix 6 §16. It was early enough in the year to encounter the 'frozen sea'; it is attractive to think that it was in early March and before the spring equinox, and that he returned to the Shetlands for the spring equinox and another *gnomon* reading, before setting off south. He would in any case almost certainly need to put in there again to stock up with food and water. We may put his final departure from the Shetlands soon after the spring equinox; he would be anxious to continue his voyage. Going south, he could go more or less directly to John o'Groats, so he would cover about 170 miles, 270 km, to reach the mainland. Perhaps with an overnight sailing he would need four days.

(8) Down the east coast of Britain and to the Calais area

Parts of the east coast in his time are now lost to the sea, which means that some of his route is now open sea; but without materially affecting distances. From John o'Groats to Dover he would cover some 1,070 miles, 1,710 km. Compared to the west coast, he would find it a more straightforward sail; here too we may allow for one or two short stopovers. Taking this into account, we can think of him reaching the Dover area about the third or fourth week in April. Another day would get him to the French coast, about 25 miles, 40 km. Note that Dion (1966) 208 with fig II, p 207, repeated *id* (1977) 194–5, proposed that from the Northern Isles he sailed south-east past the south-west coast of Norway into the Baltic, and returned the same way, presumably to verify the length of the east coast of Britain; this was partially adopted by Hawkes (1977) 31–8, with maps 8 and 9, pp 28, 30, who proposed that he sailed from the Northern Isles to Iceland, from there direct to the Baltic, and from there back to the Northern Isles. Dion's route is theoretically possible; Hawkes' would be if one accepted that Thule was Iceland, but both are too imaginative for Pytheas' world: he lacked the charts of previous explorers to embark on such open sea voyages with no knowledge of where any particular landmass might be; and if he had gone directly to the Baltic, it also raises the question whether measuring Britain's east coast was so important to him as to be worth the time and effort in returning to the Northern Isles or at least the John o'Groats area to do so.

His estimate was 15,000 stades, 1,700 miles, 2,720 km (F4). Here, the discrepancy is larger than that suggested for the west coast (see above): a round figure of 12,000 stades (1,360 miles, 2,175 km) would be more realistic. Perhaps he overestimated his progress for this stretch, finding it easier sailing than up the west coast, and believing that he had gone further each day than he had. He also overestimated the south coast, though that can be explained by the basis on which he noted it (p 176).

(9) From Calais-Dunkirk to the Baltic

It is doubtful that he reached as far as the Vistula; the analysis in Appendix 4 suggests that he got no further than the Rostock area, or a little further east. Even so, his average rate of progress would be modest; there are considerable stretches where the coastal shape is irregular, often with offshore islands. Once past Skagen he would not necessarily find the shortest distance south as he negotiated the islands that comprise modern Denmark, especially as they are tricky without local knowledge: see on F13 §55. A fair estimate is that he would need to cover 1,450 miles, 2,320 km, to reach the Rostock area, and once there he would need time to sail around to find out what he could about amber. He clearly spent a few days on shore on his amber island. If we allow him two months to reach the area, an average of only 25 miles, 40 km, per day, we can envisage his arriving at the end of June and leaving in the middle of July. He would be somewhere in the Denmark area at the summer solstice, consistent with F39, that he took some sort of measurement in that area.

(10) From the Baltic back to Calais-Dunkirk

Another 1,450 miles, 2,320 km, on the return journey he would be able to make slightly better time, with the incentive to get home, and his knowledge of the coasts from his outward journey; even so, he would need some five or six weeks, and reach the Calais area about the second week of August.[15]

(11) From Calais-Dunkirk to Corbilo

This stretch would require him to cover some 875 miles, 1,400 km. He would need three to four weeks, passing Pointe St Mathieu about the end of August. At the point where he had left the French coast to cross the English Channel on his outward journey, so roughly 600 miles, 960 km, from Calais, he would assume that the English coast opposite was the same length; he overestimated at 7,500 stades, 850 miles, 1,360 km: see on F4. After Pointe St Matthieu, he was passing coasts familiar from his outward voyage, though he would not want to sail overnight for this stretch. We may put him at Corbilo about the second week of September.

(12) From Corbilo to Marseille

From here homewards he could use his familiarity with the coasts from his outward voyage to speed up his return with some overnight sailings. From this point onwards his crew would be anxious to make good time and get back home. There would now be some 7 to 8 weeks to the end of the conventional sailing season, the end of October, or a few days into November. The distances for this stretch,

---

15 The distance has been put as the same as the outward journey, though it may well have been a little less (and need less time), as he would be able to take a more direct course going out of the Danish islands.

(1) to (3) above, total 2,955 miles, 4,730 km; it is assumed that, as on the outward journey, he kept to the coast and did not attempt to cross the Bay of Biscay. Sailing only by day, he would need an average of some 50 to 65 miles, 80 to 105 km, per day; but there would be stretches where overnight sailings were feasible, and, as suggested, undertaken. He would be just back in Marseille at the end of the sailing season. The autumn equinox would occur around the time he approached or was passing the modern frontier between France and Spain on the Atlantic coast.

If the above timetable is thought too tight a schedule, we can suggest that he overwintered in the Jutland area and got home in the third year. His finite resources and the likely attitude of his crew tips the balance of probability towards two rather than three years.[16]

6 Pytheas arrives home

When he arrived home, he wrote about his voyage, *Peri tou Okeanou* 'About the Ocean': F3, F7. Notwithstanding the generic '*World Guide*' mention in F22, and the 'Cadiz to Tanais' exaggeration in F30, where the problems are discussed, the sense of all our references to him, and to others relying on him, is that he wrote just one book, about the unknown Ocean and the north of Europe, and with no supplement (or second book) about the Mediterranean.[17] Enough survives to enable us to suggest in broad terms where he went and what he did. Map 4a shows his journey on a modern map, with the names for places and other features he noted or may have noted. Map 4b repeats the names on a revised map 1, p 9.

On the basis of what survives, map 4b shows how Pytheas perceived the parts of Europe after his voyage; what had changed and what had not.

Sections 4 and 5 above show that it is not difficult to propose a route, though, as variously noted, others propose both a different route and a much longer voyage. That of itself tells us how much we have lost; the unevenness in our references makes it hard to assess what sort of details Pytheas did or did not include in his book. Where we have no evidence that he mentioned something, is it because it was not in his book, or it was, but we have lost it? To some extent that involves the separate question whether he kept a log or notes during his voyage; we face the same question in relation to Herodotos' travels and Xenophon's *Anabasis*. Whether or not his ship had a cabin, his luggage could have included a chest in which to store papyrus and wax tablets; but we cannot feel confident that he did that, and we must be careful not to adopt our standpoint; we take written records

16 The unrealistic notion that has been floated that he abandoned his boat in the Baltic and returned home via an overland journey to the Black Sea has been dismissed in the Note at the end of F30.
17 For example, when Strabo complains that Eratosthenes or Hipparchos relied on Pytheas' lies, he did not need to specify which book had the lies. On this point Posidonios' similar title p 2 is neutral.

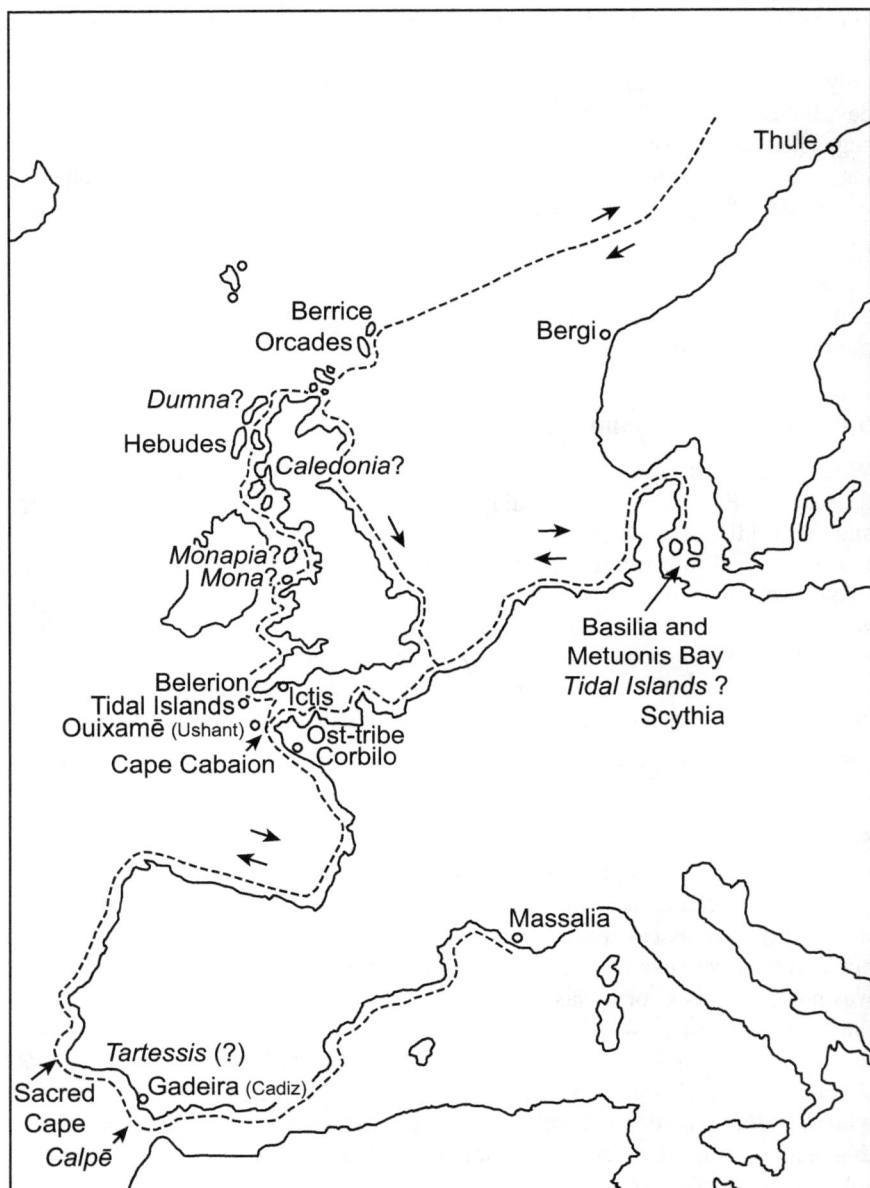

Map 4a Modern map showing Pytheas' voyage. Names and places which he noted, or we can be reasonably confident he noted, are in normal type; those he may have noted are in italics plus '?'.

Source: Drawn by the author, adapted from www.deviantart.com/eddsworldbatboy1/art/ Blank-Europe-map-324717588

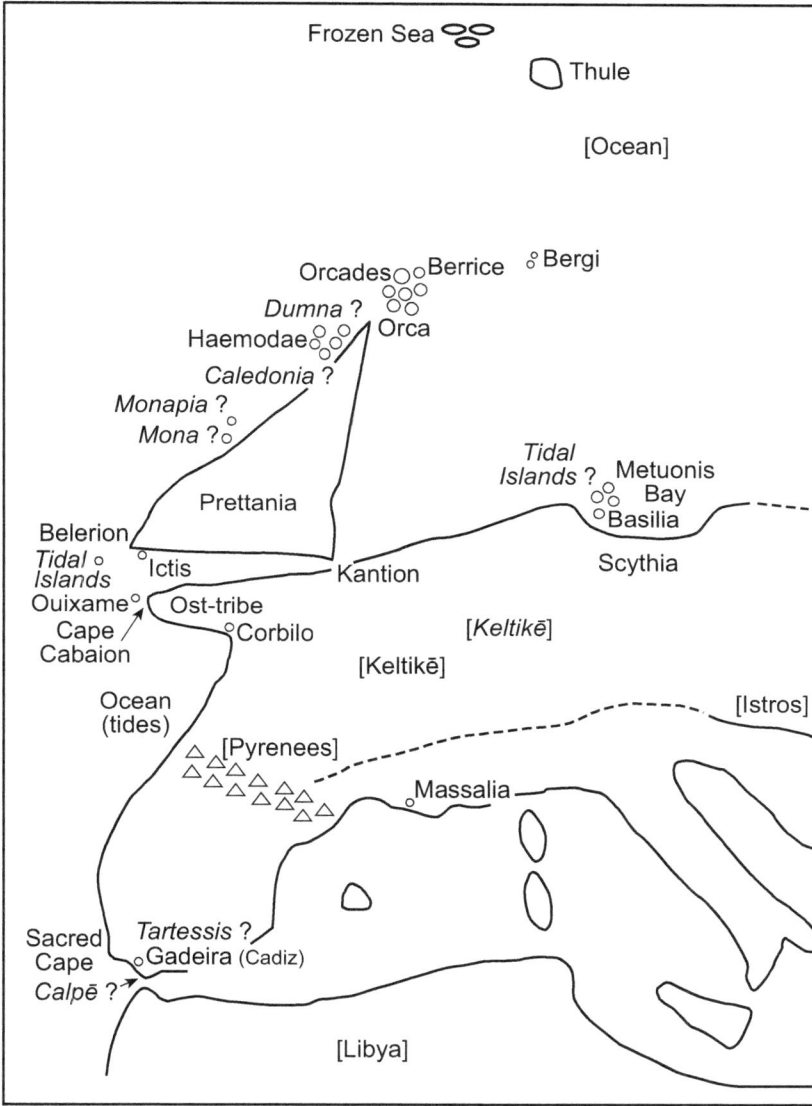

Map 4b Sketch map showing Pytheas' view of northern Europe after his voyage, showing how that had changed in the light of what he had learnt, as recoverable from
surviving references. It may be compared to map 1, p 9. Names and places which
he noted, or we can be reasonably confident he noted, are in normal type; those
he may have noted are in italics plus '?'. Names in square brackets are for other
features of the world which he would have believed existed but would probably
not have featured in his book.

Source: Drawn by the author.

for granted. Greeks are correctly said to have had better memories than us; they had to, living in a world where so much information was transmitted orally; but the question whether his book was written from memory or from notes is unanswerable. We must not be seduced by thinking of nineteenth century explorers, who could easily take notebooks and pencils.

Looking first at astronomy, as discussed in Appendix 2 §§2–4, if he took *gnomon* readings as he went, he would need not only to note them, but add enough detail of his location to make the readings meaningful for others. Assuming that he did, and Hipparchos' trigonometry enabled him to turn equinox and winter readings into summer solstice ones and hours of daylight, the locations proposed here, and the daylight figures which would correspond to his readings, would be: first year summer solstice, north coast of Iberia, around Gijón, 15½ hours; autumn equinox, Portpatrick area, 17 hours; winter solstice and second year spring equinox, Lerwick area, 19 hours; summer solstice, Copenhagen area, 17½ hours; autumn equinox, around San Sebastián, 15½ hours. Whether Hipparchos did use Pytheas' figures (if any) as a cross-check against his own calculations must remain an open question (see on F28).[18] F38 and F39 suggest that at least one piece of astronomy was in the Baltic.

He made several references to waves and tides, important to a mariner. He noted them at Cape St Vincent, F34, and once in the Atlantic his observations over the next few weeks enabled the scientist in him to deduce that the moon was a major factor in causing the significant tides he encountered, different from those at home (F1). He noted tidal islands off Cornwall, probably the Scillies or perhaps the Channel Islands, F5, as well as the waves off the Scottish mainland, F16, and the spring tides in the Baltic, F21. It is feasible that he was the source behind Mela's tidal islands discussed on F13 and Appendix 4 §§5–8.

Both the possible locations for *gnomon* readings and his noting waves and tides are relevant when we consider how methodically or otherwise he recorded geography. He could well have noted that he passed Calpē, Gibraltar. It was a definite milestone; he was passing from home waters, the Inner Sea, into the Ocean. We have his five days from Cadiz to Cape St Vincent, F34, the first stage of his voyage into 'new' waters, since up to Cadiz there would be no need to say anything special. Anything else he said before reaching coast opposite the northern end of the Pyrenees is lost. His surprise at how long it took him to circumnavigate Iberia, F34, is as consistent with noting the moon's phases as keeping a log. As there noted, we cannot assume that he recorded the distance he had covered, and so we cannot know if Eratosthenes could use him for dimensions of Iberia when calculating of the length of the world, criticised by Strabo in F26. We next know

18 This point is equally valid if one proposes an alternative and longer voyage; the locations for his *gnomon* readings (apart from Marseille) are still absent from our references.

Pytheas at Corbilo on the Nantes estuary, where he learnt that the island called Prettania was four days away. After Cadiz, Corbilo is the only port, harbour, or trading station which we can be certain he named. It was another significant point in his voyage; he learnt that there was a large island to explore before proceeding further along the Ocean coast of Europe. As he crossed over to Britain, he noted Cape Cabaion, whether Pointe du Raz or Pointe de St Matthieu, and the offshore island of Ushant, Ouixamē, F26. If he recorded the other significant cape on this part of his voyage, Cabo Touriñán, Cape Nerion, at the north-west corner of Iberia, it has not survived. From Cape St Vincent up to arriving in Cornwall, we have no more than F26 and F34 to F37. It is impossible to know what 'lies' Pytheas recorded that Strabo complains of; we cannot say how Pytheas described Cape Cabaion that Strabo said was wrong, F37.

A good deal survives about Britain. As noted above, he mentioned a group of tidal islands somewhere off Cornwall. He learned and noted its three promontories, Land's End (Belerion), Dunnet Head or Duncansby Head (Orca), and the Dover area (Kantion); he deduced that it was triangular, and recorded the lengths of its three sides with such accuracy as he could (F4). He noted that Britain was visible from the Calais area, F4; but if he was the source for later geographers stating that her south coast as a whole was opposite the coast of Keltikē, as Mela, F12, and Strabo, Appendix 7, that does not survive. As he sailed north, he may have learnt the name Caledonia: see on F18. He noted the Northern Isles, the Orcades, and the Hebrides, the Hebudes, and perhaps some others (F19); on arrival he had noted Ictis, St Michael's Mount (F4 with F19). The mariner in him noted the unusual small boats, the coracles (F19), and his general curiosity noted some local features: the hospitality in Cornwall, F4, and the diet elsewhere, F38. According to Strabo, he said that Kent was 'several days' from Keltikē; this problematic statement is noted on F25. In the north, he recorded that Thule was six days further north, and the 'frozen sea' another day's sail away: Appendix 6 §§2, 4, 7, 8, and §§14, 16.

The difficult question is what, if anything, he said about Ireland. There is no text which connects him to Ireland, either for its name, its location, or any details of its people and way of life. He would not be aware of Ireland as he sailed up the west coast of Britain. Ireland is too far away to be visible as he passed Bardsey Island and then Anglesey. After that, he would have the Isle of Man on his port side. But further north he would be able to glimpse Antrim, the eastern part of which is mountainous. As he passed where Portpatrick and the Killantrigan lighthouse are now, it was some 22 miles distant; the huge rocky promontory of Fair Head is only 14 miles away from where the Mull of Kintyre lighthouse now stands. But in terms of autopsy, he would see only that there was a small hilly island to the west or south-west. On the other hand, there are two arguments to suggest that he might have learnt something about Ireland. One is that as he sailed up the west coast of Britain, he could have learnt about Ireland from traders; the Fishguard area and Anglesey suggest themselves as possible places for this sort of information. The other is that when he next put in to land after the Mull of Kintyre, he might ask about the small island he had just seen. If he recorded Mona and Monapia, he would presumably want to record this one as well.

Two scholars in particular have looked at classical references to Ireland. So far as Pytheas is concerned, Tierney (1976) 259–60 is particularly enthusiastic; he argues that virtually everything in later writers about Ireland comes from Pytheas: 'although Pytheas may not have visited Ireland . . . it is virtually certain . . . that he obtained at least hearsay knowledge'. He refers to statements about the shape of the island, that it lies north of Britain, its rich pasture, and the savagery and incest and cannibalism of the Irish, and argues that they come from Pytheas, citing Strabo 4.5.4, Diodoros 5.32, and Mela 3.53.[19] At 264–5, he argues that Pytheas was one of Ptolemy's sources for the names of capes with sailing distances between them to help calculate the distances between them. A more sober assessment by Freeman (2001): Pytheas must have seen the peaks of the Antrim mountains, but is cautious about including Ireland in Strabo's and Polybios' use of 'British Isles' (33–4); and Caesar's information on Ireland 'probably derives from earlier Greek geographers such as Pytheas' (37), which means it could equally have come from someone else. He notes each reference to Ireland by Diodoros, Strabo, and Mela, and nowhere suggests that any of it comes from Pytheas. The balance of probability is that Tierney was wrong. If Pytheas had recorded anything about Ireland, Pliny, who had his book, would have noted it, and we would have expected to find at least the name of the island in his list in F19. We may also question whether a man of scientific bent like Pytheas would record lurid stories about incest and cannibalism; he was not an ethnographer even if naturally curious. There were clearly many merchants' and traders' stories about Ireland that circulated later; but we cannot go further than to say that if Pytheas did record the name of a small island west of the Mull of Kintyre, or even if he heard about contact with Ireland when in (say) the Fishguard area, and noted it, we have lost it.

From Britain, he sailed into the Baltic, but little detail survives either for this stage, or his return home. We have the Metuonis Bay, recorded as 6,000 stades long, F21, and the island where amber is washed up whose name he noted, probably, as Basilia: see Appendix 4. He said something about tides here; Pliny referred to an 'aestuarium', F21, 'tidal coast'. Mela, F13, had sources which led him to believe that there were two separate groups of tidal islands in the Baltic, although in fact there is only one, modern Denmark. Although by his time various reports about the north would have filtered through to the Mediterranean, it is tempting

---

19 He could have added Strabo 2.1.13 and 2.5.8, that the Irish are miserable savages living in a cold climate; though he is there stressing the cold, as Ireland is at the northern limit of the habitable zone. Cf p 8 n 5.

to think that at least some of Mela goes back to Pytheas' autopsy: see Appendix 4 §8. But, as with Iberia and the Atlantic coast of France, we are left to speculate. Pytheas certainly appears to have said rather more than survives. On the one hand, Strabo remarks several times that Pytheas told lies about the Ocean coast of northern Europe (F26, F31, F35, F38, F39), but chose not to indicate what the lies were. Indeed, he asserted that the land beyond the Elbe was unknown (7.2.4). Strabo would not want to record stories of men with horses' feet, as Mela, F13, and Pliny, F17, did; but Pytheas said something about the area, and Strabo ignored him. It is ironic that Marcianus thought of both Pytheas and Strabo as good geographers: see on F9 with n 3. But that also does not help on specifics.

That effectively sums up what we can say about his book with reasonable certainty. It is frustrating that we can comment on what he did not say at such length. However he described the Metuonis Bay and a tidal coast and tidal islands in that area, he did not leave a clear indication that Jutland existed as a large peninsula. Neither Mela nor Strabo knew of it (F13, Note to Appendix 4 §3); Pliny knew of it because of the Roman expedition round it noted on F17. There are no place names such as ports, apart from Cadiz and Corbilo; even if normally anchoring offshore at nightfall, he would need to put in periodically for food and water, and learn the name for where he was. If he took *gnomon* readings, he would have had to locate them, but we have no names. There are no river names; apart from Britain, we could have expected the Rhine and perhaps the Elbe. On the other hand, he did not call his book a *Periplus*, and we cannot say that he wanted to write a mariner's guide to the Ocean coast of Europe, with the names of the harbours and other aids to navigation, as noted on F9; the absence of Cape Nerion, for instance, has been noted above.

Beyond Britain we have no other islands, the Frisians in particular, or any of the islands either side of Jutland other than Basilia. He noted the Ost- tribe in Brittany, Appendix 5, but no tribes in Britain. Although he found tin in Cornwall, F5, he does not appear to have said anything to rebut the existence of the tin islands (pp 6, 8–9). Our ability to create a long list of what is not there does not help on whether he kept notes, nor whether they were never there in the first place. It shows only how little of his book survives. For all we know, Pytheas finished his book with a brief sentence to the effect that having found his amber island, he returned home, and his only observation of note was to see how long it took him to sail from the French coast opposite Kantion, the Dover area, and reach the point he judged to be opposite Belerion, Land's End, to fill in the dimensions of his British triangle. Any review of the references to him leaves us frustrated, with the many unanswerable questions as to how much we have lost.

CONCORDANCE

Scott	Mette	Roseman	Bianchetti
F1	F2	T26	F2a, F2b
F2	F14	T27	F12a
F3	F9(b)	F9	F13b
F4	-	-	-
F5	-	-	-
F6	-	-	-
F7	9(a)	F8	F13a
F8	F1	F1	F1
F9	F3	T28	F23
F10	F13(a)	T18(b)	F9b
F11	-	T29	F12b
F12	-	-	-
F13	-	-	-
F14	F10(a), 10(b) T24	T17, T20, F17	F3, F7d, F10, F17
F15	F13(a)	T18(a)	F9a
F16	F13(a)	T19	F7e
F17	F11(b)	T21	F16
F18	F11(b)	T22	F7c
F19	F11(b)	T23	F8f
F20	F13(b)	-	-
F21	F11(a)	T25	F15
F22	F15	T30	F19
F23	F6(f)	T1b	F18b
F24	F6(a)	F2	F8a
F25	F6(a)	T1a, T2, F3	F7a, F8a, F18a
F26	F6(a)	T3, F5	F6a
F27	F6(b)	T4	F8b
F28	F6(b)	T5	F11
F29	F5	T6	F20
F30	F7(a)	T7, T8	F7b, F8d, F21

Scott	Mette	Roseman	Bianchetti
F30 n 8	F7(b)	-	-
F31	F6(c)	T9, F6	F8c
F32	F6(c)	-	-
F33	F6(d)	T10	F14
F34	F8	T11	F4
F35	F4	T12	F35
F36	F7(c)	T13	F5
F37	F6(e)	T14	F6b
F38	F6(g)	T15, F7	F8e
F39	F6(h)	T16	F8g

ABBREVIATIONS

Barrington   *Barrington Atlas of the Greek and Roman World* (Princeton and
              Oxford 2000, ed Talbert R J A)
BNP   *Brill's New Pauly*, ed Schneider H et al (Leiden 2002-, also online)
*CAH*III.3   *The Cambridge Ancient History*, 2nd edn, vol III part 3 (Cambridge
              1982)
CIL   *Corpus Inscriptionum Latinarum*, vol XIII, ed. O Hirschfeld (Berlin
              1899, repr 1966; also available online)
CGL   The Cambridge Greek Lexicon, ed Diggle J *et al* (Cambridge 2021)
GGM   *Geographi Graeci Minores*, ed Müller C F W (Paris 1855)
IG   XII.5 *Inscriptiones Graecae: Inscriptiones Cycladum* (Berlin
              1903–9)
LSJ   *A Greek-English Lexicon*, Liddell H G and Scott R, revised and aug-
              mented by Jones Sir H S with the assistance of McKenzie R and
              many other scholars, with supplement (Oxford 1996)
OCD   *The Oxford Classical Dictionary*, ed Hornblower S and Spawforth A
              (3rd edn Oxford 1996; 4th edn Oxford 2012)
OLD   *Oxford Latin Dictionary*, ed Glare P G W (Oxford 2012)

BIBLIOGRAPHY

Arnaud P *Les routes de la navigation antique* (Paris 2005)

Asheri D *et al A Commentary on Herodotus Books I–IV* (Oxford 2007)

Aubet M E *Mainake: The Legend and New Archaeological Evidence*, in Osborne and Cunliffe (2005) 187–202

Aubet Semmler M E a *Phoenician Trade in the West: Balance and Perspectives*, in Bierling (2001a) 97–112

Aubet Semmler M E b *Some Questions Regarding the Tartessian Orientalizing Period*, in Bierling (2001b) 225–40

Aujac G *Introduction aux Phénomènes, Geminos* (Paris 1975)

Badian E *A King's Notebooks* HSPh 72 (1967) 183–204

Beaujeu J *Pline L'Ancien, Histoire Naturelle, livre II* (Paris 1950)

Belén Deamos M *Phoenicians in Tartessos*, in Dietler and López Ruiz (2009) 193–227

Berthelot A *Festus Avienus, Ora Maritima* (Paris 1934)

Bianchetti S *Pitea di Massilia, L'Oceano* (Pisa and Roma 1998)

Bianchetti S *The 'Invention' of Geography: Eratosthenes of Cyrene*, in Bianchett *et al* (2015) 132–45

Bianchetti S, Cataudella M R and Gehrke H-J (eds) *Brill's Companion to Ancient Geography* (Leiden 2015)

Bierling M R (ed and translator) with Gitin S *The Phoenicians in Spain* (Winona Lake, IN 2001)

Boardman J *The Greek Overseas* (4th edn London 1999)

Bowen A C and Goldstein B R *Hipparchus' Treatment of Greek Astronomy* Proc Am Phil Soc 135 (1991) 233–54

Bowen A C and Todd R B *Cleomedes' Lectures on Astronomy* (Berkeley and Los Angeles 2004)

Bowen E G *Britain and the Western Seaways* (London 1972)

Braund D and Wilkins J, eds *Athenaeus and his World* (Exeter 2000)

Bridgman T P *Hyperboreans: Myth and History in Celtic-Hellenic Contacts* (New York and London 2005)

Broggiato M *I Frammenti/Cratete di Mallo* (La Spezia c 2001)

Buck C D *The Greek Dialects* (Chicago 1955)

Carpenter R *Beyond the Pillars of Hercules* (London 1973)

Cary M and Warmington E *The Ancient Explorers* (London 1929)

Casson L *Travel in the Ancient World* (Baltimore and London 1994)

Casson L *Ships and Seamanship in the Ancient World* (Baltimore and London 1995)

Cautadella M R *Eudoxus of Cnidus and Dicaearchus of Messena*, in Bianchetti *et al* (2015) 115–31

Celestino S and López Ruiz C *Tartessos and the Phoenicians in Iberia* (Oxford 2016)

Champion T C and Collis J R, eds *The Iron Age in Britain and Ireland: Recent Trends* (Sheffield 1996)

Chantraine P *Dictionnaire étymologique de la langue grecque* (Paris 1968–1980)

Churchill Temple E *The Templed Promontories of the Ancient Mediterranean* The Geographical Review XVII (3) (1927) 353–86

Cunliffe B *The Extraordinary Voyage of Pytheas the Greek* (London 2001)

Cunliffe B *On the Ocean: The Mediterranean and the Atlantic from Prehistory to AD 1500* (Oxford 2017)

Cunliffe B *The Ancient Celts* (2nd edn Oxford 2018)

Dalby A *Siren Feasts* (London and New York 1996)

Denniston J D *The Greek Particles* (2nd edn Oxford 1954)

Detlefsen D *Der Kenntniss der Alten von der Nordsee* Hermes 32 (1897) 192–201

Dicks D R *The Geographical Fragments of Hipparchus* (London 1960)

Dicks D R *Solstices, Equinoxes, and the Pre-Socratics* JHS 86 (1966) 26–40

Dicks D R *Early Greek Astronomy to Aristotle* (London 1970)

Dietler M and López Ruiz C *Colonial Encounters in Ancient Iberia* (Chicago 2009)

Dilke O A W *Greek and Roman Maps* (Baltimore and London 1985, 1998)

Diller A *The Textual Tradition of Strabo's Geography* (Amsterdam 1975)

Dion R *La renomméee de Pythéas* REL 43 (1965) 443–66

Dion R *Pythéas explorateur* RPh 40 (1966) 191–216

Dion R *Aspects politiques de la géographie antique* (Paris 1977)

Dunbabin T J *The Western Greeks* (Oxford 1948)

Dunbar N *Aristophanes, Birds* (Oxford 1995)

Edelstein L and Kidd I G *Posidonius, Vol I The Fragments* (2nd edn Cambridge 1989): see Kidd I G, *below*

Evans J and Berggren J L *Geminos' Introduction to the Phenomena* (Oxford 2006)Finkelberg M *The Geography of the Prometheus Vinctus* Rh Mus 141 (1998) 119–141

Freeman P *Ireland and the Classical World* (Austin 2001)

Frick G *Pomponius Mela, De Chorographiai* (Lipsiae 1888)

Frisk H *Griechisches Etymologisches Wörterbuch* (Heidelberg 1972)

Gantz T *Early Greek Myths* (2 vols, Baltimore and London 1996)

Gasparotto G *Geometria: De Nuptiis Philologiae et Mercurii, liber sextus* (Verona 1983)

Gibbs S L *Greek and Roman Sundials* (New Haven 1976)

Gilespie C C *Dictionary of Scientific Bibliography vol 1* (New York 1980)

Gilula D *Hermippus and his Catalogue of Goods*, in Harvey and Wilkins (2000) 75–90

Gisinger F *Die Erdbeschreibung des Eudoxos von Knidos* (Leipzig and Berlin 1921)

Goldstein B R and Bowen A C A *New View of Greek Astronomy* Isis 74–3 (1983) 330–340

Gomme A W A *Historical Commentary on Thucydides vol I* (Oxford 1945); *vol IV* (1970)

González Ponce F J *Sobre el valor histórico atribuile al continedo del Ora Maritima* Faventia 15 (1) (1993) 45–60 (ddd.uab.cat/pub/faventia/02107570v15n1/02107570v15n1p45.pdf)

Gonzalez Poncé F J *Avieno y el Periplo* (Ecija 1995)

Gørstad H, Skar B, og Skre D *Historien I forhistorien, festkrift til Einar Østmo på 60-års dagen* (*Kulturhistrik museum Skrifter 4*) (Oslo 2006)

Greenidge A H J and Clay A M *Sources for Roman History, 133–70 BC* (2nd ed, rev Gray E W (Oxford 1960)

Grenfell B P and Hunt A S *The Hibeh Papyri Part I* (London 1906)

Grønvik O *Thule – det eldste navnet på landet vårt?*, in Gørstad *et al* (2006)

Guthrie W K C *A History of Greek Philosophy I, II* (Cambridge 1962, 1965)

Hammond N G L *Alexander the Great* (London 1980)

Hammond N G L *A History of Greece* (3rd edn Oxford 1986)

Hansen M H and Nielsen T H *An Inventory of Archaic and Classical Poleis* (Oxford 2004)

Harley J B and Woodward D (eds) *The History of Cartography* vol 1 chapter 8 (*prepared from materials supplied by Germaine Aujac*) (Chicago 1987)

Harvey D and Wilkins J, eds *The Rivals of Aristophanes* (London 2000)

Hawkes C F C *Europe and the Greek Explorers* (Oxford 1977)

Heath Sir Thomas *A History of Greek Mathematics* (2 vols, Oxford 1921; repr with corrections New York (1981)

Herzog R *Restauration und Erneuerung: die lateinische Literatur von 284 bis 374 n. Chr* München 1989

Heubeck A and Hoekstra A, *A Commentary on Homer's Odyssey, Vol II* (Oxford 1989)

Hornblower S *A Commentary on Thucydides vol I (Oxford 1991); vol III* (Oxford 2008)

Hunt D W S *An Archaeological Survey of the Classical Antiquities of the Island of Chios Carried out between the Months of March and July 1938* BSA 41 (1940–1945) 29–52

Jacob A *Curiae Strabonianae* RPh (2e série) 36 (1912) 148–78

Jay M and Richards M P *British Iron Age Diet: Stable Isotopes and Other Evidence* Proceedings of the Prehistoric Society 73 (2007) 169–190

Jones M *Plant Exploitation*, in Champion and Collis (1996)

Joorde Raffael (2016): www.academia.edu/27786980

Kannicht R *Tragicorum Graecorum Fragmenta vol 5.1* (Göttingen 2004)

Keyser P and Irby-Massie G (eds) *Encyclopedia of Ancient Natural Scientists* (London 2008)

Kidd I G *Posidonius Vol II, The Commentary* (2 vols, Cambridge 1988); (*Vol I, The Fragments, Edelstein L and Kidd I G* (2nd edn Cambridge 1989); (*Vol III, The Translation of the Fragments* (Cambridge 1999))

Kirk G S and Raven J E *The Presocratic Philosophers* (Cambridge 1960)

Lasserre F *Ostiéens et Ostimniens chez Pythéas*, Mus Helv 20 (1963) 107–13

Lasserre F *Die Fragmente des Eudoxos von Knidos* (Berlin 1966a)

Lasserre F *Strabon Géographie tome II* (Paris 1966b)

Lasserre F *Strabon Géographie tome III* (Paris 1967)

Lloyd A B *Herodotus Book II* (Leiden 1976; 2nd edn 1994)

López Ruiz C *Tartish and Tartessos Revisited*, in Dietler and López Ruiz (2016) 255–280

Luraghi N (ed) *The Historian's Craft in the Age of Herodotus* (Oxford 2001)

Malkin I *Religion and Colonization in Ancient Greece* (Leiden 1987)

Mata D R a *The Ancient Phoenicians of the 8th and 7th Centuries BC*, in Bierling (2001) 155–198

Mata D R b *The Beginnings of the Phoenician Presence in Southwestern Andalusia*, in Bierling (2001) 263–298

Matthews J *Continuity in a Roman Family: The Rufii Festi of Volsinii*, Hist 16 (1967) 484–509

McPhail C *Pytheas of Massilia's Route of Travel*, Phoenix LXVIII (2014) 247–57

Mette H J *Pytheas von Massalia* (Berlin 1952)

Most G W (ed) *Collecting Fragments, Fragmente Sammeln* (Göttingen 1997)

Müllenhoff K *Deutsche Altertumskunde, Erster Band* (Berlin 1870)

Murphy J P *Avienus: Ora Maritima* (Chicago 1977)

Nansen F, trans Chater A G *In Northern Mists vol 1* (New York 1911)

Nash D *Reconstructing Posidonios' Celtic Ethnology: Some Considerations* Britannia 7 (1976) 111–26

Ogilvie R M and Richmond I A *Tacitus: Agricola* (Oxford 1967)

Olmos R *Los Griegos en Tartessos: Una Nueva Construction Entre las Fuentes Arqueológicas y las Literarias*, in Aubet Semmler (1989) 495–519

Osborne R and Cunliffe B, eds *Mediterranean Urbanisation, 800–600 BC* (Oxford 2005)

Østmo E *The History of the Norvegr, 2000 BC–1000 AD*, in Skre (2020) 3–66

Pelling C *Fun with Fragments*, in Braund and Wilkins (2000) 171–190

Pothecary S *Strabo, Polybios, and the Stade* Phoenix 49 (1995) 49–67

Radt S *Strabons Geographika* (Göttingen 2002–2011)

Rhodes P J *A Commentary on the Aristotelian Athenaion Politeia* (Oxford 1993)

Rivet A L F and Smith C *The Place Names of Roman Britain* (London 1979)

Robertson D S *The Evidence for Greek Timekeeping* CR 54 (1940) 180–2

Roller D W *Through the Pillars of Herakles: Greco-Roman Exploration of the Atlantic* (London 2006)

Roller D W *Eratosthenes' Geography* (Princeton and Oxford 2010)

Roller D W *A Historical and Topographical Guide to the Geography of Strabo* (Cambridge 2018)

Romer F E *Pomponius Mela's Description of the World* (Ann Arbor 1998)

Roseman C H *Pytheas of Massalia On the Ocean* (Chicago 1994)

Rouillard P *Greeks and the Iberian Peninsula: Forms of Exchange and Settlements*, in Dietler M and López Ruiz (Chicago 2009) 131–51

Schmitt R *Medisches und Persisches Sprachgut bei Herodot* ZDMF 117 (1967) 119–45

Schulten A *Avieno, Ora Maritima* (Barcelona and Berlin 1922; 2nd edn Barcelona 1955)

Scott L *Historical Commentary on Herodotus Book 6* (Leiden 2005)

Silberman A *Chorographie, Pomponius Mela* (Paris 1988)

Skre D (ed) *Rulership in 1st to 14th Century Scandinavia* (Ergänzungsbände zum Reallexicon der Germanischen Altertumskunde, Band 114) (Berlin and Boston 2020)

Smolak K *Avienus*, in Herzog (1989) 320–7

Sommerstein A *Aristophanes Ecclesiazousae* (Warminster 1998)Stahl W H, Johnson R, and Burge E L *Martianus and the Seven Liberal Arts* (translation) (New York 1977)

Ste-Croix G E M de *The Origins of the Peloponnesian War* (London 1972)

Stichtenoth D *Ora Maritima* (Darmstadt 1968)

Svennung J *Scadinavia und Scandia* (Uppsala 1963)

Theiler W *Posidonios Die Fragmente* (Berlin and New York 1982)

Thomas R *Oral Tradition and Written Record in Classical Athens* (Cambridge 1989)

Thomas R *Literacy and Orality in Ancient Greece* (Cambridge 1992)

Thomas R *Herodotus' Histories and the Floating Gap*, in Luraghi (2001) 198–201

Thomas R *Polis Histories, Collective Memories and the Greek World* (Cambridge 2019)

Thompson D'A W *A Glossary of Greek Fishes* (Oxford 1947)

Thomson J O *History of Ancient Geography* (Cambridge 1948)

Tierney J J *The Celtic Ethnology of Posidonius*, Proc Royal Ir Acad 60 (1959–60) 189–275

Tierney J J *The Greek Geographical Tradition and Ptolemy's Evidence for Ireland*, Proc Royal Ir Acad 76 (1976) 257–265

Tierney J J and Bieler L *Dicuil: Liber de Mensura Orbis Terrae* (Dublin 1967)

Toomer C J *Apollonius of Perga*, in Gillespie (1970) 179–193

Toomer C J *Ptolemy's Almagest* (London 1984)

Ussher R G *Aristophanes, Ecclesiazousae* (Oxford 1973)

Walbank F W *Polybius* (Berkely and London 1972)

Walbank F W *Commentary on Polybius* vol I (Oxford 1956, repr 1970), vol III (Oxford 1979)

Windstedt E O *The Christian Topography of Cosmas Indopleustes* (Cambridge 1909)

Zehnacker H and Silberman A *Pline l'ancient, Histoire Naturelle Livre IV* (Paris 2015)

LIST OF PASSAGES CITED

(other than the fragments themselves)

Greek

Agathemeros 1 *86 n 1*
Alcman *PMG* 90 *11 n 20*
Anacreon *PMG* 361 *14 n 2*
Anaxagoras 59 A1 D–K, 59 A42 D–K
    41 n 4
Anon *PMG* 894 *110 n 15*
Apollonios Paradoxographus 24
    Giannini *10 n 14*
Apollonios Rhodios
    IV 282–6 *17 n 21*
    IV 327 *69 n 26*
    IV 505–6, 580 *67 n 16*
Aratos 497–9 *21 n 8*
Aristophanes
    Eccl 652 *39 n 4*
    Frogs 475 *16*
    F169 *135 n 8*
    F227 *135 n 9*
    F680 *111 n 16*
    F695 *39 n 4*
Aristotle
    Ath Pol 67.3–4 *39 n 4*
    De Caelo
    285b 25–30, 296b 24–298b 21,
      297a 8–b 23, 297b 23–31 *20 n 5*
    297a–298a *44*
    298a 16–18 *20*
    De Mundo 383b *63*
    Marvels
    844a 24–30 *141*

846b 29–30 *144 n 13*
    Meteor
    346b 21–347a 12, 354b 24–33 *160 n 20*
    350b, 362b *11 n 21*
    360b 2 *108*
    361b 36–362a 31 *160 n 20*
    362a 32–b 9 *20, 28*
    363a 21–365a 13 *43*
    365a 32 *20 n 5*
    366a 19 *25 n 4*
    PA 581a 18 *97*
Arrian *An*
    7.1.1–4 *167*
    7.15.4–6 *168 n 6*
    7.16.1–3 *167 n 3*
    7.23.2 *18 n 18*
Athenaios
    2.61d *118 n 11*
    4.151e–152f *99 n 10*
    4.152c *119 n 14*
    4.153e *99 n 10*

Bacchylides *Epinic* 3.58–62 *11 n 17*
Baton *Murderer 137 n 13*

Cassius Dio
    55.28 *61*
    80.20–23 *47 n 1*
Cleomedes
    I.1.195–208, I.3.22–43 *28 n 6*
    I.4.30–43, 184–96 *27 n 3*
    II.1.438–44 *54*

LIST OF PASSAGES CITED

4.152 *14, 107*
4.192.3 *16*
4.196 *13*
5.31.2 *110*
6.21.1 *135, 169 n 11*
8.6.1, 9.101.2 *39 n 3*
Hesiod *Theog*
285–94, 979–83 *108*
293 *108 n 6*
Op
171 *110 n 15*
504–8 *11 n 19*
F150 M–W *11 n 16*
Hipparchos *Comm in Arat*
1.2.18, 1.2.20 *28 n 5*
1.2.22, 1.3.9–10 *20*
1.8.6 *41*
1.10.13–15, 1.10.17 *28 n 5*
Homer *Il*
8.14 *25*
14.246 *10 n 12*
15.171 *11 n 19*
18.607–8 *10 n 12*
19.358 *11 n 19*
Od
10.79–85 *2 n 3*
10.80–132, 12.4 *10 n 13*
Hym Hom *Dion* 29 *16 n 11*

Iordanes *Get III.16–IV.25* *60 n 13, 147 n 18*
Isidoros of Seville
Etym 14.6.5, 33 *65*
Orig 11.3.19 *59 n 10*

Lucian *Lex* 4 *136 n 10*

Menander F625 *39 n 3*

Orphica *Argonautica* 1081–2 *69 n 26*

Pausanias 6.19.4 *109*
Philochoros *FGrH* 328 F122 *135 n 9*

Pindar *O* 3.16 *11 n 16, n 17*
P 10.30, *I* 6.23 *11 n 16*
Plato
Phaido 111e–112a *159*
Rep 548b8–c1 *43 n 3*
Plutarch
Alex 68 1–2 *167*
Mar 11.3 *144 n 11*
Mor 938d *10 n 12*
Pollux 9.46 *135 n 8*
Polybios
1.42.3–7 *30 n 2*
3.22–5 *15 n 6*
3.39.5–7 *171*
3.57.3 *30 n 1*
12.27.4 *18 n 20, 96 n 2*
34.6.15 *94 n 4, 99 n 8*
34.9.2 *110 n 12*
34.9.4 *110 n 13*
34.15.7 *96 n 1*
Posidonios
T19, T25–6 *99 n 10*
F49 *2 n 4, 106 n 1, 142 n 7, 144 n 11*
F67, F73 *99 n 9*
F206 *28 n 7*
F214–F220 *2, 25 n 5*
F219 *144 n 12*
F227b *99 n 10*
F239 *2*
F272 *142 n 8, 144 n 11*
F276 *47 n 3*
F277b *144 n 10*
Priscianus Lydus *Solutiones* VI 72.2–12 *144 n 12*
Ps–Galen *Hist Phil* 66 *24 n 1*
Ps–Scylax
21.9 *67*
68.23, 70.2 *144 n 9*
Ps–Scymnos
112–114 *21*
162–6 *16 n 10*
374 *67*

194

Latin

Orosius 1.76–9 *66 n 66*
Ovid *Metam* 10.229 *125 n 7*

Plautus
    Boeotia F1.2 *39 n 6*
    Cas 963 *19 n 1*
Pliny *NH*
    1.2, 4, 5 *70 n 2*
    1.6 *70 n 2*
    2.167 *61*
    2.169 *13*
    2.177–9 *45*
    2.182 *137 n 15*
    2.187 *134 n 3*
    2.246 *61, 68, 84 n 7*
    3.7–11 *109*
    3.11 *109 n 10*
    3.38 *125 n 7*
    3.152 *68 n 18*
    4.18 *53 n 2*
    4.41, 49, 80–1, 91 *61*
    4.61, 65 *125 n 7*
    4.82 *125 n 7*
    4.99–100 *60, 146*
    4.100 *149 n 2*
    4.107 *47 n 3*
    4.108 *113*
    4.119 *9 n 9, 68 n 21*
    4.120 *108, 109 n 8*
    5.59 *78 n 10*
    5.132 *125 n 7*
    6.183, 191, 194 *78 n 10*
    6.200, 7.155 *59 n 9*
    7.197 *9 n 9, 13, 68 n 21*
    7.214 *137 n 15*
    10.97, 12, 82, 87–93 *78 n 5*
    17.47 *113*
    29.10, 22–3 *19 n 1*
    37.31–51 *71*
    37.31 *37*
    37.32 *68 n 18*
    37.32, 40 *48 n 4*
    37.33, 36 *xx*
    37.33 *71 n 2, 73 n 8*

    37.42–3 *71*
    37.42 *67*
    37.43–5 *37, 71 n 3*
    37.45 *36 n 1*
    37.61 *143*
    37.188–9 *71 n 1*
Pliny the Younger *Ep* 3.5.3–6, 7–17 *53*

Quintus Curtius 10.1.17–19 *167*

Res Gestae Divi Augusti *(RG)* 26 *61, 62*
Rutilius *De Reditu Suo* 1.499 *66 n 6*

Seneca *Medea* 379 *3, 52, 68*
    Q.N. 5.18 *167 n 4*
Servius *Virg Georg* 1.30 *68 n 19*

Tacitus *Ag*
    4.2 *19 n 1*
    10.4 *68 n 19*
    29.2 *4 n 11*
    29–38 *27 n 4*
    Ann
    2.62 *147 n 18*
    Germ
    2, 37, 40, 41–2, 44 *146 n 18*
    2 *149 n 2*
    43 *60*
    45.1 *2 n 5*
Terrence *Eun* 341 *39 n 6*
    Phorm 514 *39 n 6*

Varro *RR* 1.2.4 *3*
Velleius Paterculus 2.104–8 *61*
Virgil *Georgics* 1.30 *3, 52, 68 n 19*
Vitruvius *De Arch*
    1.6.6 *134 n 3*
    9.7.1–2 *137 n 15*
    9.8.1 *136*
    9.8.2 *137 n 13*

Inscriptions

    CIL XIII 1/1 490 *149*

INDEX

For Product Safety Concerns and Information please contact our EU
representative GPSR@taylorandfrancis.com
Taylor & Francis Verlag GmbH, Kaufingerstraße 24, 80331 München, Germany